# SPECIAL ISSUES ON MAGNESIUM ALLOYS

Edited by **Waldemar A. Monteiro**

**Special Issues on Magnesium Alloys**
Waldemar A. Monteiro

**Published by InTech**

Janeza Trdine 9, 51000 Rijeka, Croatia



**Publishing Process Manager** Ivana Lorkovic

**Technical Editor** Teodora Smiljanic

**Cover Designe**r InTech Design Team

Additional hard copies can be obtained from orders@intechopen.com

Special Issues on Magnesium Alloys, Edited by Waldemar A. Monteiro
p. cm.
ISBN 978-953-307-391-0

For EU product safety concerns:
IN TECH d.o.o., Prolaz Marije Krucifikse Kozulić 3, 51000 Rijeka, Croatia,
info@intechopen.com or visit our website at intechopen.com.

# Meet the editor

Bachelor and teacher of Physics, MSc (Solid State Physics), DSc (Nuclear Technology) at University of São Paulo, Brazil. Materials Engineering Specialty (Brazilian Physical Researches Center, Rio de Janeiro, Brazil); Physical Metallurgy Specialty (Max Planck Institut für Metallkunde, Stuttgart, MPI, Germany). His expertise areas are Physical Metallurgy, Powder Metallurgy and Materials Characterization (Electron Microscopy and Microanalysis Techniques). He is a productivity fellowship researcher (CNPq - Brazil). He is the senior researcher in metallic materials at Nuclear and Energetic Research Institute, São Paulo, Brazil. Professor (graduate course) on Materials Characterization (Electron microscopy techniques and microanalysis). Researcher and lecturer (undergraduate and graduate courses) at Presbyterian Mackenzie University (São Paulo, Brazil). He published 2 book chapters and 140 articles (scientific journal or congress proceedings). The academic advisories include 25 scientific initiates (under graduate student), 20 MSc and 14 DrSc.

# Contents

# Preface

Magnesium is among the lightest of all the metals, and also the sixth most abundant on earth. Magnesium is ductile and the most machinable of all the metals. Magnesium alloy developments have traditionally been driven by requirements for lightweight materials to operate under increasingly demanding conditions. This has been a major factor in the extensive use of magnesium alloy castings, wrought products and also powder metallurgy components. The biggest potential market for magnesium alloys is in the automotive industry. Although significant opportunities exist for increasing magnesium alloy usage in automobiles, many of these new applications require the development of new alloys, improved manufacturing technologies and significant design and technical support for the automotive supply chain.

In recent years new magnesium alloys have been demonstrated a superior corrosion resistance for aerospace and specialty applications. Very large magnesium castings can be made, such as intermediate compressor casings for turbine engines, generator housings and canopies. Forged magnesium parts are also used in aero engine applications to be used in higher temperature applications. Other applications include electronics, sporting goods, office equipment, nuclear applications, flares, sacrificial anodes for the protection of other metals, flash photography and tools. Considering the above informations special issues on magnesium alloys are showed in this book: casting technology; density of liquid-solid Mg-Pb alloys; surface modification of some special Mg alloys; manufacturing of protective carbon coatings on magnesium alloys; fatigue cracking behaviors of cast magnesium alloys and also magnesium alloys biocompatibility as degradable implant materials.

**Prof Dr. Waldemar Alfredo Monteiro**
Science and Humanities Center
Presbyterian Mackenzie University
and
Materials Science and Technology Center
Nuclear and Energetic Researches Center
Universitu of São Paolo
São Paolo, SP
Brazil

# 1

# Casting Technology and Quality Improvement of Magnesium Alloys

By Hai Hao
*Dalian University of Technology, China*

## 1. Introduction

Magnesium alloys offer the potential for weight and related energy savings in both the automotive and aerospace industries as they have the highest strength-to-weight ratio of common structural metals. Despite higher cost, this potential benefit has lead to a recent increase in demand on cast and wrought magnesium products. This chapter will talk about numerical simulation work and technology related to magnesium casting process. And it includes four topics:

1. The fundamentally-based mathematical models to predict the temperature and stress evolution in both the billet as well as the dummy block during the DC casting of wrought magnesium alloy billets;
2. The application of EPM(Electromagnetic Processing of Materials) on the magnesium alloys;
3. High intensity ultrasonic treatment to improve the solidification structure of magnesium alloys;
4. The effects of grain refiner and the external fields on grain size and microstructure of magnesium alloys.

## 2. Research on modeling of magnesium DC casting

Direct chill (DC) casting of billets, shown schematically in Fig.1, is the main process for producing the precursor material for many nonferrous (i.e., zinc, aluminum, and magnesium) wrought products as well as the remelt stock for cast products [1]. During this process, molten metal is initially poured onto a dummy block located inside a water cooled mould. When the metal reaches a predetermined height inside the mould, the dummy block is lowered at a controlled speed. As the freshly solidified billet comes out of the mould, water is sprayed on the newly exposed surface. The DC casting process has found extensive acceptance in the light metals industry, especially for aluminum, as a reliable and economic production method, involving low capital investment, simple operating features and great product flexibility. During the 1990s, industry, especially the automotive industry, rediscover magnesium and took advantage of its remarkable properties, especially low density, to reduce weight and improve fuel economy. Magnesium also found new applications in hand tools and, most recently, in portable electronic equipment. Due to its weight-saving benefit, high mechanical properties, high damping capacity, and electromagnetic shielding, increasing markets of magnesium components are, resulting in the need for greater scientific and technical understanding of magnesium and magnesium casting process.

Although the DC casting process has been the subject of scientific study since its beginning in the 1930s and has been used almost exclusively to produce aluminum ingots/billets and more recently magnesium billets, there is still work necessary to optimize the design of the casting process from the standpoint of productivity, cost effectiveness, and final ingot quality. One of the challenges in optimization is the complex interaction between the casting parameters, such as withdrawal rate, water flow rate, dummy block design, and defect formation, which is difficult to rationalize experimentally. One approach overcome this problem is to use fundamentally based mathematical models to analyze defect formation such as hot tearing, cold cracking, bleed outs, and cold cracking, bleed outs, and cold shuts because most are directly related to heat flow and deformation phenomena. While this trend is growing in popularity, it hinges on the ability to predict the temperature evolution and subsequent thermal stress during the casting process. Over the years, computer modeling has provided a powerful means to investigate and understand the evolution of thermal and mechanical phenomena during the DC casting process.

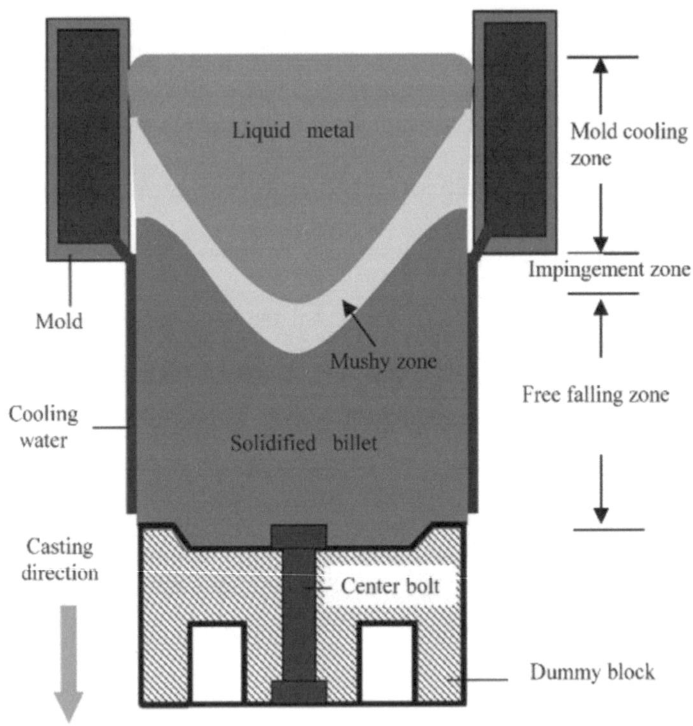


Fig. 1. Schematic DC casting process used for magnesium billet casting

Mathematical modeling of the DC casting process using various techniques has been underway since 1940[2]. The earliest published mathematical model of DC casting using a computer to solve the heat conduction equation numerically was published by Adenis et al.

in 1962[3]. In this work, the steady-state temperature distribution in DC casting magnesium alloy billets was calculated using a two-dimensional (2-D) axsymmetric heat-transfer model with heat-transfer coefficient boundary conditions. The heat-transfer coefficient boundary conditions described the separation between the billet and mold during primary cooling and the contact with water below the mold. Following this early work, interest appears of articles have been published on aluminum DC casting since the 1970s compared with only a few articles on magnesium.

The other research on modeling magnesium DC casting, besides Adenis et.al, was published by Hibbins[4]. In this work, 2-D axisymmetric and three-dimensional (3-D) steady-state heat-transfer models were developed for DC casting of AZ31 magnesium alloy using the finite-difference numerical method. The models were used to predict the steady-state temperature profiles in billets and blooms under variety of casting conditions. A series of constant heat-transfer coefficient boundary conditions, calculated based on experimental temperature measurements, were used to describe the primary and secondary cooling regions. Specifically, the primary cooling heat-transfer coefficients were reduced from 35000 to 300 $W/m^2/K$ at fixed positions within the mold to reflect the air gap formation. Constant heat-transfer coefficients of 10,000 to 12,000 $W/m^2/K$ were defined in the impingement and free falling sections of the secondary cooling region. The resulting model predictions show good agreement across different casting conditions, but the use of constant heat-transfer boundary conditions and fixed length heat-transfer zones in the primary and secondary cooling regions limits the applicability of this model to a wide range of casting conditions.

A review of the published literature on aluminum DC casting reveals that early modeling efforts adapted Adenis et al.'s work to describe the 2-D steady-state temperature distribution in aluminum alloy billets[2,5]. However, heat-transfer models with increased sophistication soon followed including those that considered transient heat conduction, 3-D geometry and complex boundary conditions. Within this published body of work, some of these models have ignored the presence of the bottom block, choosing to describe the interfacial heat transport along the base of the ingot as a heat-transfer coefficient boundary condition with a fixed far-field temperature, while others have included the dummy or bottom block and have described the interfacial heat transfer between the billet and bottom block as a function of base deformation, which is assumed to evolve during casting[6].

In terms of secondary cooling, boiling water heat transport has typically been described by an effective heat-transfer coefficient, which is a nonlinear function of surface temperature and vertical position on the billet surface[7]. Other advancements include correlations that are a function of water flow rate and impingement point temperature an include the effect of water ejection[8]. The most recent thermal models of aluminum DC casting have been coupled with fluid flow and deformation models to understand and describe inter-related transport phenomena, such as water incursion below the base of the ingot[9].

Another trend that is emerging in the aluminum DC casting literature related to hot tearing and include not only the development of criteria and models to predict the onset of hot tearing but their implementation within DC casting thermal/stress models. A recent publication by Drezet and Rappaz et al. has successfully used a pressure based hot tear model to predict hot tearing in the center of aluminum billets in proximity to the base[10].

Typical defects that occur during DC casting include both hot tearing and cold cracking that can lead to downstream defects during subsequent processing operations, and are major difficulties which restrict the productivity of the process and its variability of alloys and ingot size. Investigations focusing on hot tearing indicate that these tears are likely generated when thermally induced stresses are applied to regions of the billet that are at or above the solidus temperature. During billet casting, metal starts solidifying from the outer surface to the center of the billet because the outer surface is being cooled by the cooling water. After the outer shell has contracted upon freezing, the inner metal tries to contract as it freezes. Because of the difference in the contraction from the surface to the center of the billet due to the differences in temperature gradient, internal thermal stress development. The temperature gradient, internal thermal stresses develop. The internal stress cause hot tears when these stresses exceed the flow stress limit of the alloy being cast. On the other hand, the stress may persist in the billet even though in the absence of temperature gradient, which is called residual stress, and can cause cold cracking. Overall, understanding the evolution of thermal stress is a prerequisite to solve the cracking problems.

In Hao's work, a previously developed axisymmetric model describing the evolution of temperature during the DC casting of magnesium AZ31 billets has been extended to predict the evolution of stress and strain in order to predict the susceptibility of the process to hot tearing using the Rappaz-Drezet-Gremaud(RDG ) criterion[11]. The as-cast constitutive behavior of the AZ31 alloy was established from compression experiments made using a Gleeble 3500 thermo literature. Residual strains/stresses on an as-cast billet combined with process deformation data can provide the data necessary to validate the mechanical model.

### 2.1 Measurement of residual strains in magnesium billets

As mentioned above, DC cast products experience thermal strains because of the shrinkage during the casting process. The thermal strains result in residual stresses after final cooling to ambient temperature. While it is difficult to investigate the stress state of the hot strand during the casting process, the residual stress state of the cold material can be analyzed by several experimental methods, classified as fully destructive, partly and non-destructive techniques.

The partially or fully destructive, or so called 'mechanical' techniques generally involve removal of material by drilling, cutting, slitting or sectioning combined with strain gauge measurements[12-13]. These methods give bulk residual strain values, but suffer the disadvantages of involving destruction of the component, and are usually limited to symmetrical components to avoid uncertainties.

The non-destructive or so called 'physical methods', mainly include ultrasonic and diffraction technique. The former method uses the electromagnetic-acoustic transducer as an 'ultrasonic strain gauge' to measure the strains. The diffraction techniques usually utilize X-rays or neutrons to measure strain states and are suitable for investigating specimens made from polycrystalline materials. Both X-ray and neutron diffraction methods measure directly changes in the lattice spacing of crystals then obtain the strain components. Residual stresses are then calculated from these strains in a similar way to those from strain gauge readings. The X-ray technique is now well known but X-rays interact with orbiting electrons and are strongly absorbed after penetrating a very small depth in most metals, making them suitable for the measurement of surface strains but not for bulk measurements. The more recent technique of neutron diffraction enables non-destructive internal strain measurements to be made, sometimes at depths of several centimeters due to the great penetrating power of the neutrons.

Since neutrons have wavelike properties they can be diffracted by the scattering object that has length scales comparable to the neutron beam wavelength. Taking advantage of the weak interaction of the uncharged neutron with electrons, which allows them to penetrate several centimeters into most metals, the neutron diffraction technique provides means to investigate bulk strains in metal components[14-15].

Hao et al. [16] presented the strain distributions along radial, axial and hoop directions in a direct chill cast billet of AZ31 magnesium alloy by neutron diffraction, which provide the data necessary to validate a thermo-mechanical model that predicts the evolution of stress/strain during the DC casting and subsequently to investigate the cracking defects in the billets. Schematic view of the billet orientation for radial strain measurement with respect to the incident and diffracted beams is shown in Fig.2.

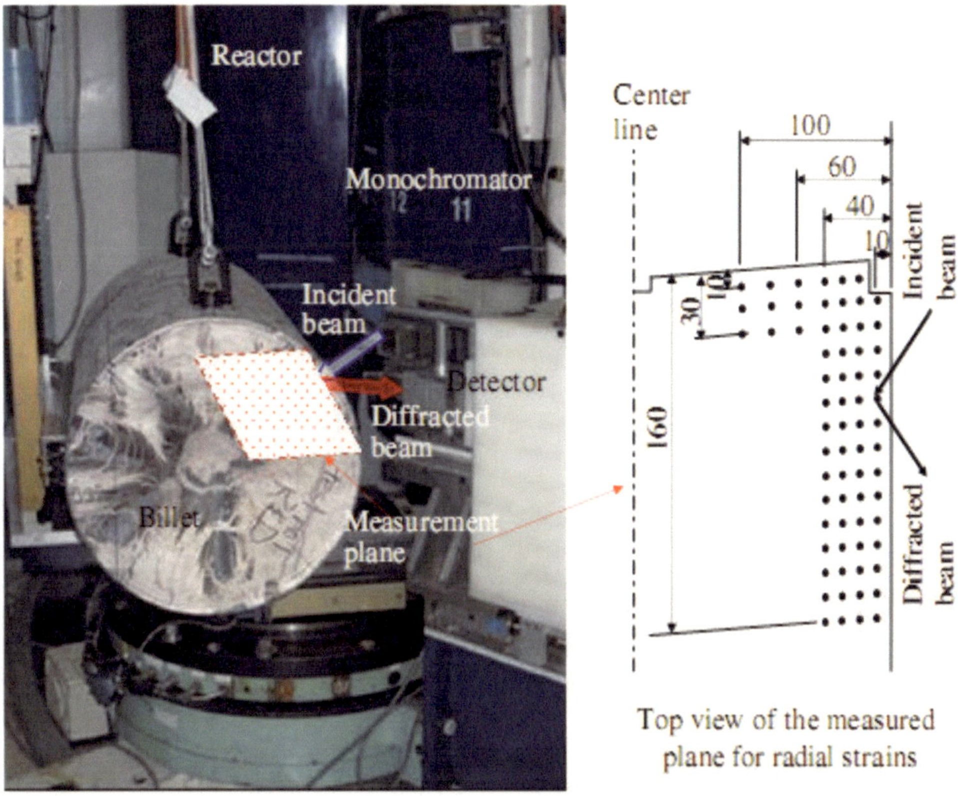


Fig. 2. Neutron diffraction apparatus and the schematic beam locations of radial strain measurement

Fig.3~5 display the strains components measured by neutron diffraction, based on the strain measurement results, the stress component can be calculated by using the following equation:

$$\sigma_i = \frac{E}{1+v}\left[\varepsilon_i + \frac{v}{1-2v}(\varepsilon_a + \varepsilon_r + \varepsilon_h)\right] \quad (1)$$

Where $\sigma_i$ and $\varepsilon_i$ are the stress and strain, respectively, in one of the three directions, E the Young's modulus and $v$ the poisson ration for the measured specimen. This information could be used to estimate the cracking tendency in the direct chill cast AZ31 billet.

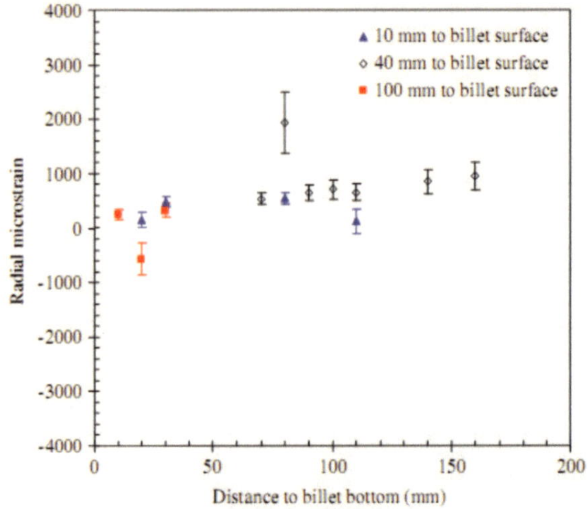


Fig. 3. Measured radial strain at different paths (10, 40 and 100 mm to the billet surface)

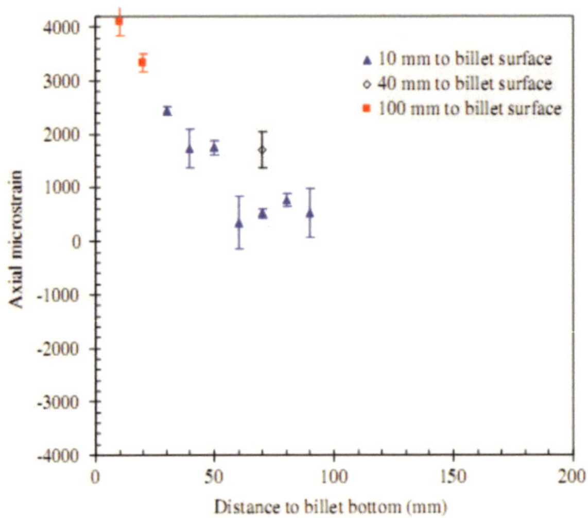


Fig. 4. Measured axial strain at different paths (10, 40 and 100 mm to the billet surface)

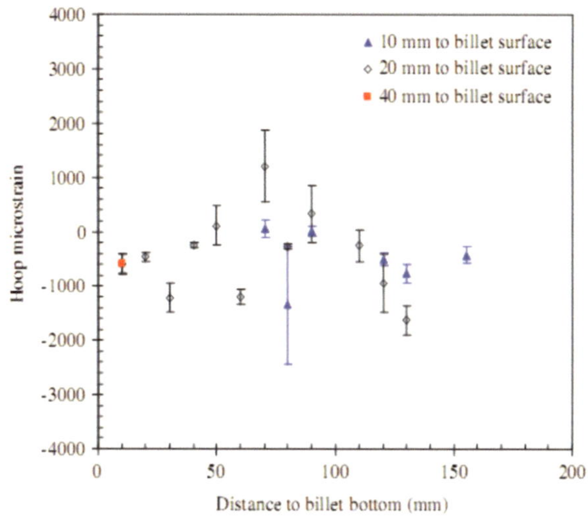


Fig. 5. Measured hoop strain at different paths (10, 20 and 40 mm to the billet surface)

**2.2 Modeling the stress-strain behavior during DC casting of magnesium billets**

Mathematical modeling of the DC casting process has been the focus of study from the middle of the twentieth century. However, since Adenis et al.[3] reported their modeling work on DC casting of magnesium in 1962, very little other work has been done on this alloy system, and to date, no attempts to predict stresses, strains, or hot tearing during magnesium alloy DC casting have been reported. In contrast, a considerable body of work has been reported on modeling the DC casting process in aluminum alloys. The most recent of these include the majority of the relevant phenomena that are thought to affect heat transfer and stress/strain development: boiling water heat transfer during cooling, water incursion between the base of the ingot and the bottom block, and macroscopic ingot distortions (butt curl and lateral pull in). There have also been some attempts to integrate various hot tearing criterions, which incorporate mushy zone pressure drop and strain-rate effects, appears to be the most successful in qualitatively predicting the correct location of hot tearing in DC cast billets. Alternative approaches to predict hot tearing include a strain-based criterion [17] or a stress-based criterion [18]. While significant progress has been made, fully quantitative hot tearing predictions remain elusive, in part due to the stochastic nature of this defect.

H.Hao et al. [19] reported their work on modeling the stress-strain behavior and hot tearing during DC casting of AZ31. In this work, a coupled thermal-mechanical axisymmetric simulation of the DC casting process for magnesium AZ31 cylindrical billets has been developed using the commercial FE package ABAQUS. The model domain section of this geometry was included in the model. A schematic of the model domain is shown in Fig.6, for a billet with a length corresponding to 505 seconds of casting time. The domain consists of 3582 elements and 3833 nodes, each approximately 10 mm×10 mm in size. All three parts of the domain—billet, dummy block, and the center bolt—were modeled using four noded isoparameteric coupled temperature/displacement elements. To simulate the casting process, a Lagrangian approach was used, whereby the thermal boundary conditions describing the primary and secondary cooling regions were moved up along the domain at a rate consistent with the billet elements were incrementally added based on the mold filling rate and casting speed.

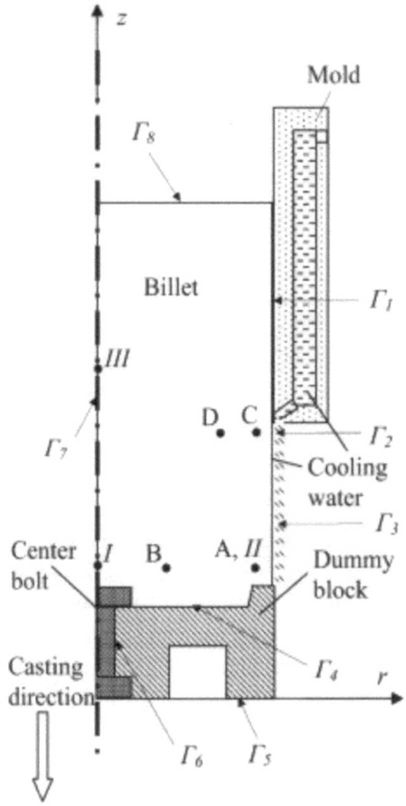


Fig. 6. Schematic of DC casting process used for magnesium billet casting, calculation domain, the relevant surfaces (Γ1 through Γ8 ) for application of boundary conditions, the thermocouple locations (A through D) during the plant trial, and the points ( I through III) of interest for the stress-stain analysis.

The governing partial differential equation for the transient thermal analysis in cylindrical coordinates is

$$\frac{1}{r}\frac{\partial}{\partial r}\left(k(T)r\frac{\partial T}{\partial r}\right)+\frac{\partial}{\partial z}\left(k(T)r\frac{\partial T}{\partial z}\right)=\rho c_p\frac{\partial T}{\partial t} \quad (2)$$

Where $r$ and $z$ are the radial and axial directions in meters, respectively; $k$ Is the thermal conductivity in W $m^{-1}$ $K^{-1}$; T is the temperature in Kevin; $\rho$ is the density in kg $m^{-3}$; and $c_p$ is the specific heat in J $kg^{-1}$ $K^{-1}$. The latent heat released during solidification is incorporated into Eq.[2] by modifying the specific heat term for temperatures within the solidification interval according to $c_{p.e}=c_p+L\frac{df_s}{dT}$ ,where $c_{p.e}$ is the equivalent specific heat, $L$ is the latent heat of fusion in J $kg^{-1}$ ,and $\frac{df_s}{dT}$ represents the rate of change of fraction solid with temperature. In the mechanical analysis, the stress and strain increments are derived based on the nodal displacements along with the compatibility and constitutive equations. The resulting total strain vector, $\Delta\varepsilon^{tot}$, is given by

$$\Delta\varepsilon^{tot} = \Delta\varepsilon^{el} + \Delta\varepsilon^{th} + \Delta\varepsilon^{pl} \tag{3}$$

Where $\Delta\varepsilon^{e}$is the elastic strain increment, i$\Delta\varepsilon^{th}$ s the thermal strain increment, and $\Delta\varepsilon^{pl}$ is the plastic strain increment. Note that the constitutive equation is based on an elastic/rate-independent plastic material formulation.

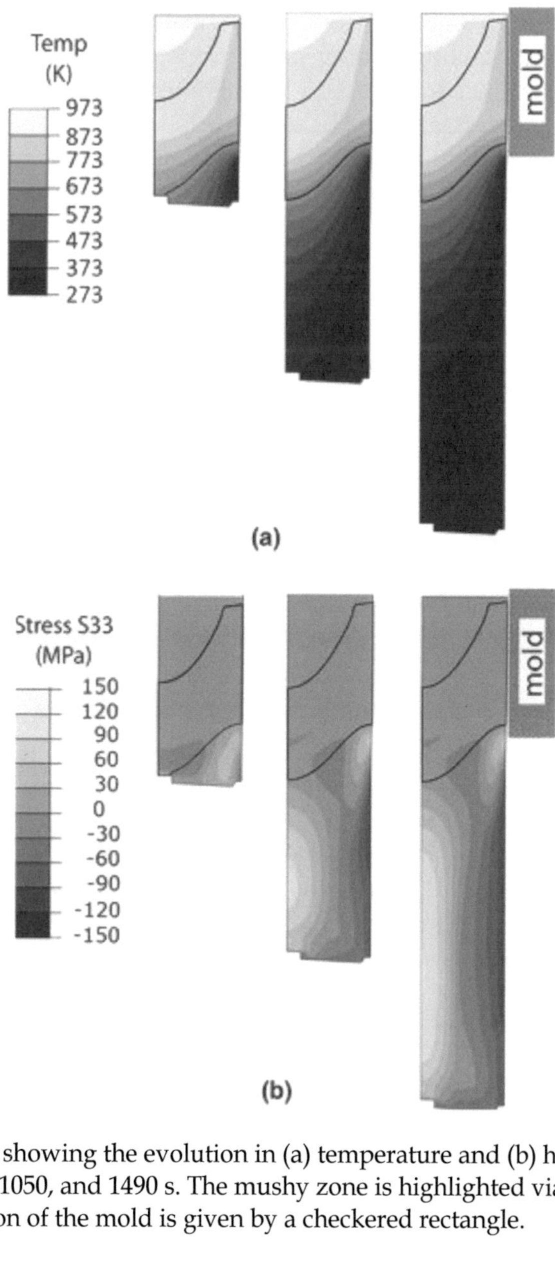


Fig. 7. Contour plots showing the evolution in (a) temperature and (b) hoop stress predicted by the model at 505, 1050, and 1490 s. The mushy zone is highlighted via a black contour line, while the location of the mold is given by a checkered rectangle.

Fig.7 shows contour plots of temperature and hoop stress in the cross section of the billet after 505, 1050, and 1490 seconds. The hoop stress is shown since, per Eq. $\dot{\varepsilon}_{\perp}^{pl} = \dot{\varepsilon}_{r}^{pl} \sin\gamma - \dot{\varepsilon}_{z}^{pl} \cos\gamma + \dot{\varepsilon}_{\theta}^{pl}$(where $\gamma$ is the angle between the thermal gradient and the radial axis $\theta$ is the hoop direction ),it is considered to be the major driving force for crack initiation and hot tear propagation in billet casting. The mushy zone has been out lined in the figures at 505, 1050, and 1490 seconds using a black line. As can be seen from the thermal contours, cooling is dominated by the secondary water cooling, which strikes the ingot surface just below the mold. Since the mushy zone does not appear to be changing size or shape relative to the mold in the three thermal contours shown, it would appear that steady-state thermal conditions are reached before 505 seconds. At the ingot center, the pool depth is estimated to be 0.2 m by 505 seconds. As shown in the contours presented in Fig.7(b), the surface of the billet below the mold is in a state of tensile stress, due to the thermal contraction induced by the cooling water sprays. Moving down the ingot, as the thermal gradient moderates, the surface stress state becomes compressive while the center region is in tension to maintain internal equilibrium. The length of the surface region in compression and the length of the center region in tension, below the water impingement zone, increase with increasing cast length. The distribution of stresses arises, because the tensile stresses that are generated at the surface of the ingot near the point of secondary cooling water impingement exceed the yield point of the material resulting in the accumulation of tensile plastic strain. Once this material cools it is placed into compression and the interior material into tension. The maximum value of the hoop stress is ~150 MPa, well above the yield point of the as-cast structure. It can also be seen that the mushy zone remains in a low state of tensile stress throughout the casting process. While this stress value is low, it has exceeded the material's yield limit resulting in permanent deformation.

The thermomechanical simulation can be used to provide a detailed description of the evolution of stress and strains during the industrial casting of magnesium alloys.

## 3. Application of EPM on DC casting of magnesium alloys

Besides the conventional casting technology, this part introduces the application of EPM (Electromagnetic Processing of Materials) on the magnesium alloys. EMC is a technology developed by a combination of MHD and casting engineering. The casting method employs the effects of electromagnetic forces upon the liquid metal placed in the alternating electromagnetic field, which is induced by an inductor. The electromagnetic forces are produced by interaction of eddy currents induced in the metal with the magnetic field of the inductor. The main advantage of the EMC technology consists in the presence of stirring motions in the melt, which lead to a significant reduction of the grain size in the solidified product. Moreover, surface quality and subsurface quality are improved due to the absence of ingot mold. The surface finish of the ingot is usually smooth enough to be hot rolled without the scalping operation that is required following direct chill casting. Besides refining internal structures, electromagnetic stirring also has advantages of homogenized alloy elements, reducing porosity and segregation, and minimizing internal cracks. Because of these distinct merits of EMC technology, many scientists and engineers in different countries are engaged in this field.

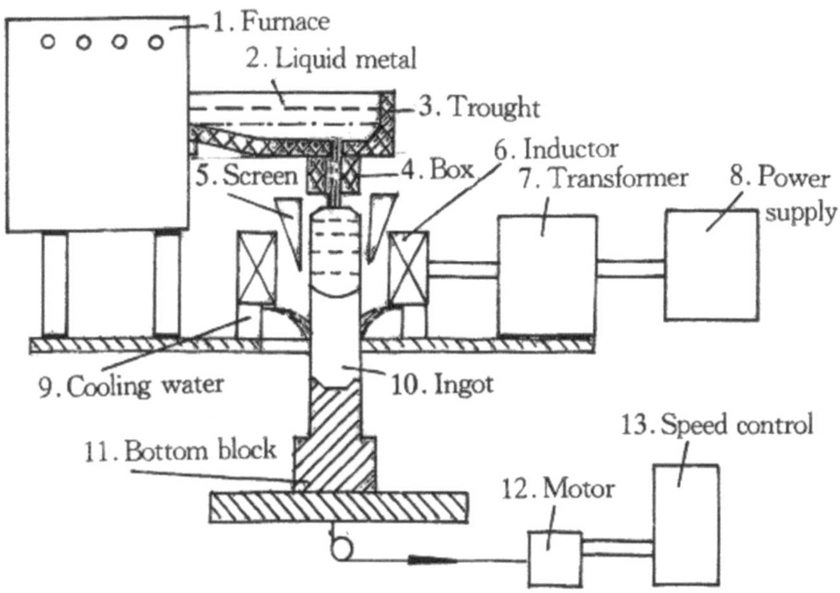


Fig. 8. Schematic diagram of EMC equipment

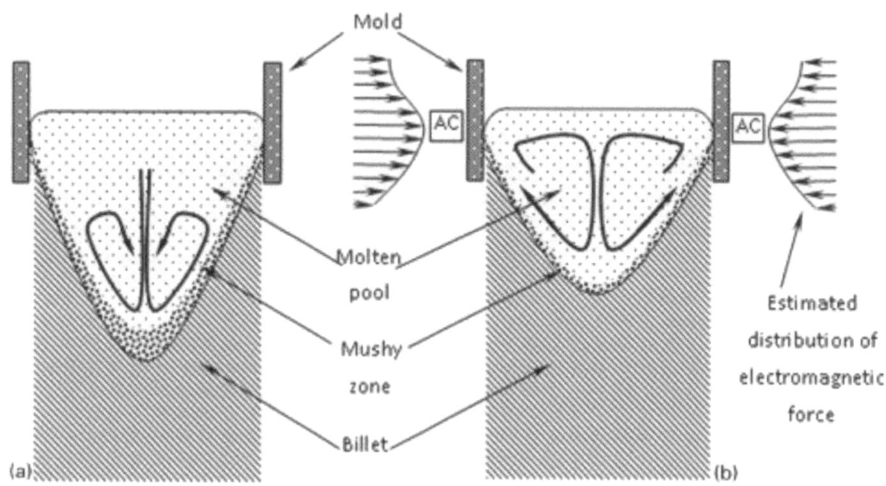


Fig. 9. Schematic diagram of a DC and b MFEMC

The continuous casting of aluminum is the foundation of the electromagnetic casting (EMC), which began from the direct chill casting invented by Aloca corporation and Vlw corporation in 1935[20]. The principle of EMC was firstly described by Getselev and his co-workers in 1960[21]. And then, they cast the first EMC ingot in laboratory in 1966. Thereafter, the industry-scale ingots with diameter from 200mm to 500mm were cast in 1969. Subsequently, this method was spread to the former Czechoslovakia and other Eastern

European countries. The principal advantage of the technology is that the metal is cast without contacting a physical mold depending on the electromagnetic forces, which excludes liquation build-ups and feather, and consequently, the surface finish of the ingot is usually smooth enough to be hot rolled without scalping operation. Because of the strong magnetic field, the structure and properties of the EMC ingot become much better. Since 1970's, occident has developed the technology in a big degree. The ingots of aluminum, copper, zinc, magnesium and their alloys were cast. At the same time, the new methods lying on different direction such as GE Levitation EMC and Horizontal EMC were implemented for casting ingots[22].

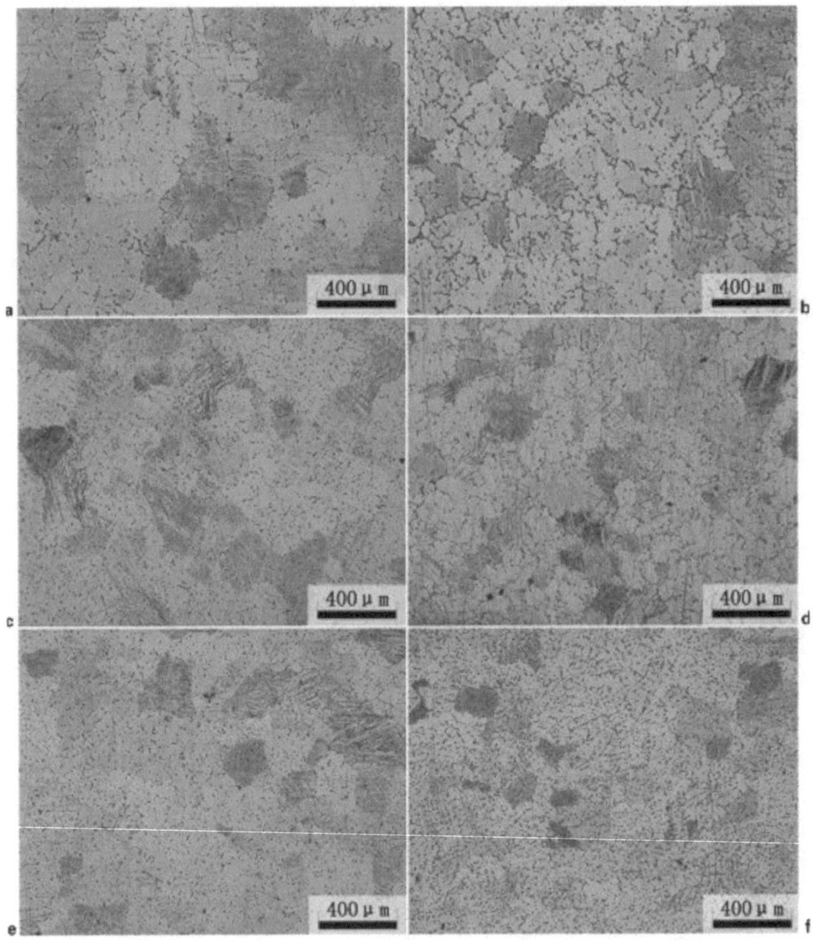


a: border of DC; b: border of MFEMC; c: one-half radius of DC; d: one-half radius of MFEMC; e: centre of DC; f: center of MFEMC;

Fig. 10. Microstructures of AZ31 alloy billets cast in different processes

The basic apparatus of EMC consists of delivery system, casting control system, shaping and cooling system, melt furnace and power supply, as shown in Fig.8[23]. The shaping system composed of an inductor, screen, cooling water box and bottom block is the main component of this piece of equipment. A medium frequency alternating current is used to generate the alternating magnetic field in the molten magnesium. This magnetic field generates a heavy eddy current on the surface of the molten magnesium in opposite phase to the imposed current through the electromagnetic coil. These results in forces directed towards the center of the ingot. The electromagnetic force located within the upper liquid part of the ingot prevents the metal from touching the mold. A metal ring screen is necessary to control the magnetic field in the top of the melt, to keep the balance between the electromagnetic pressure and the hydrostatic pressure, and to achieve optimum horizontal flow and distribution of the liquid metal(Fig.9). Recently, with the development of supper conducting magnet technology, a new branch of EPM, materials processing under a high magnetic field is dramatically highlighted. The magnetic intensity of the high magnetic field can reach $10^3$ times stronger than that of the common magnetic field. The effects of magnetic force of high magnetic field on the paramagnetic and diamagnetic materials can't be ignored any more. Many interesting phenomena have been found, such as orient alignment of the structures , variation of solid-state phase transformation, etc.

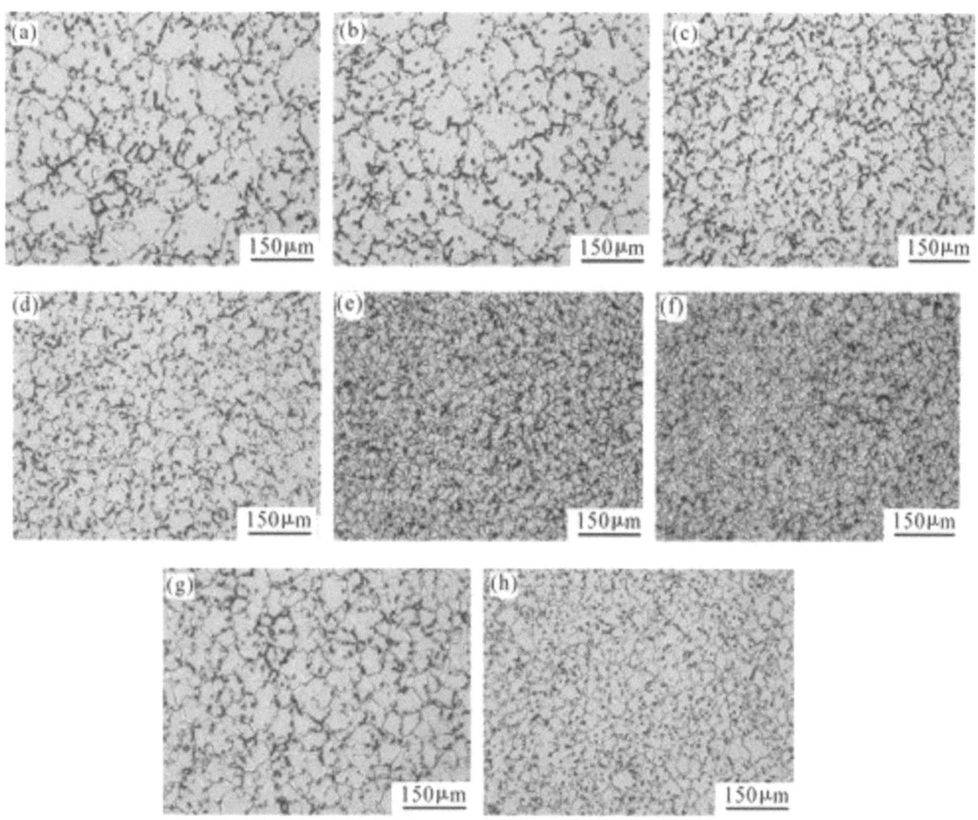


Fig. 11. Microstructure of ZK60 alloy billets cast under different electromagnetic powers:(a)DC casting edge; (b) DC casting center; (c) EMC-5KW edge; (d) EMC-5KW center;(e) EMC-10kW edge; (f) EMC-10kW center;(g) EMC-20kW edge;(h) EMC-20kW center

Billets of AZ31 magnesium alloy with and without intermediate frequency electromagnetic field were investigated by Pang et al. [24] In his work, compared with microstructures and mechanical properties of the DC casting billet, the medium-frequency electromagnetic continuous casting (MFEMC) billets shows refined and even microstructures throughout the whole section of the billet and improved mechanical properties, the microstructures of AZ31 billets cast in different processes are shown in Fig.10. Ren et al.[25] have studied the effects of middle frequency electromagnetic field on the precipitations of ZK60 magnesium alloys, the results show that the microstructure are refined and distribution uniformity of precipitations is observed after applying the middle frequency electromagnetic field(Fig.11). The refined microstructure is in connection with increased nuclei which are likely to be as a result of electromagnetic undercooling which decreases the free energy barrier of nucleation and increases the nucleation tendency by an induced undercooling $\Delta T$ and forced convection. The movements between grain sizes of different locations in the billet are a result of particles' forced movements with particles in the inner area moving outward and particles in the border area moving inward.

## 4. Effects of ultrasonic field on Mg-based alloys

Magnesium alloys are getting increased attention for their low density, high specific strength, high specific rigidity and good damping capacity. However, the use of magnesium alloys has been restricted by their limited mechanical properties. Several previous investigations proposed that high intensity ultrasonic treatment was one of the effective ways to improve the solidification structure and the mechanical properties of metals. Ultrasonic vibration of aluminum alloys had been studied extensively, and it can effectively refine the grain size. Investigations carried out between 1960 and 1990[26], mainly in the former Soviet Union countries, clearly demonstrated its grain-refining effects on magnesium alloys and significantly improved mechanical properties. The introduction of powerful ultrasonic oscillations into the melt can be quite simply adapted to the commercial technologies of continuous casting (vertical, horizontal DC casting, strip casting, etc.) and shape casting (precise, die casting, liquid forging, etc.) Ultrasonic degassing, an environmentally clean and relative inexpensive technique, should be paid more attention on speeding up the industrial application and revealing the mechanism the effects on the solidification process.

Fig.12 is the illustration of a direct ultrasonic treating process. The ultrasonic equipment is comprised of a 20 kHz ultrasonic power, an ultrasonic transducer made of piezoelectric ceramics, an ultrasonic amplitude transformer and an ultrasonic probe. The ultrasonic amplitude transformer and probe are made of stainless steel. The grain refinement of ultrasonic treatment on the microstructure of alloys is based on the physical phenomena arising out of high-intensity ultrasound propagation through the liquid. Considerable work has been carried out to determine the grain refinement mechanisms by ultrasonic treatment and two underlying mechanisms have been proposed for ultrasonic grain refinement based on cavitation: (i) cavitation-induced (shock waves) dendrite-fragmentation and (ii) cavitation-enhanced heterogeneous nucleation [27-30]. Cavitation-induced dendrite fragmentation hypothesis assumes that the shock waves generated from the collapse of bubbles lead to fragmentation of dendrites, which are redistributed through acoustic streaming and increasing the number of crystals [30-31]. Cavitation-enhanced heterogeneous nucleation interpreted further in terms of two different mechanisms. The first is the pressure pulse-melting point ($T_m$) mechanism [28-29], where the pressure pulse induced by the collapse of a bubble alters $T_m$ according to the Clapeyron equation$\Delta T_m = T\Delta P\Delta V/\Delta H$.

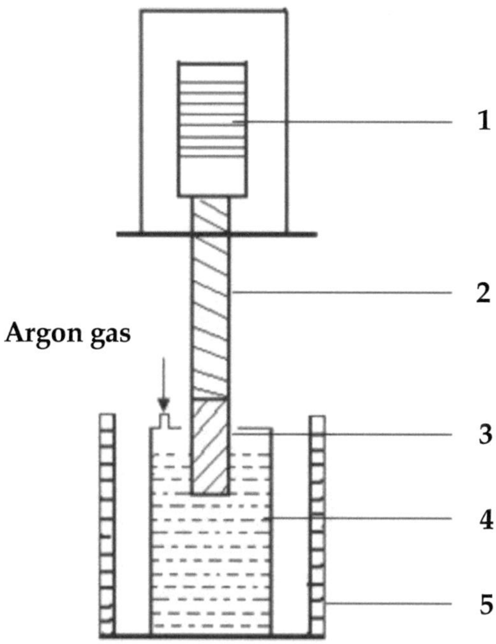


Fig. 12. Schematic diagram of the experiment apparatus 1- Ultrasonic transducer; 2- Amplitude transformer; 3-Ultrasonic probe; 4- Stainless steel mould; 5- Heat preserving furnace.

An increase in $T_m$is equivalent to increasing the undercooling and so an enhanced nucleation event is expected. The second mechanism is cavitation-induced wetting[28], where the defects (cavities or cracks) on the substrate surfaces with the pressure pulse can act as effective nucleation sites, leading to enhanced nucleation[28].

Mg-Li series alloy are called ultra-light magnesium based alloys because they are the lightest metal structural material. They have high specific strength and stiffness, good damping capacity, and electromagnetic shielding properties. It will reduce the energy consumption if Mg-Li series alloys are successfully widespread applied. But the strength of Mg-Li alloys at room temperature especially at high temperature is low, which limits their applications. In order to obtain the uniform microstructure and high strength of Mg-Li alloys, Yao et al. [32] introduced the ultrasonic vibration into the solidification process of the Mg-8Li-3Al alloy. With the effects of Ultrasonic treatment, the morphology of α phase was modified from coarse rosette-like structure to fine globular one (Fig.13), and the tensile strength and elongation were improved by 9.5% and 45.7%, respectively. With the purpose of investigating the mechanism of grain refinement under ultrasonic vibration, the effects of ultrasonic vibration power on fluid field is described by particle image velocimetry (PIV). Fig.14 shows the ultrasonic filed can transmit in the fluid and form circulation flow to uniform the microstructure.

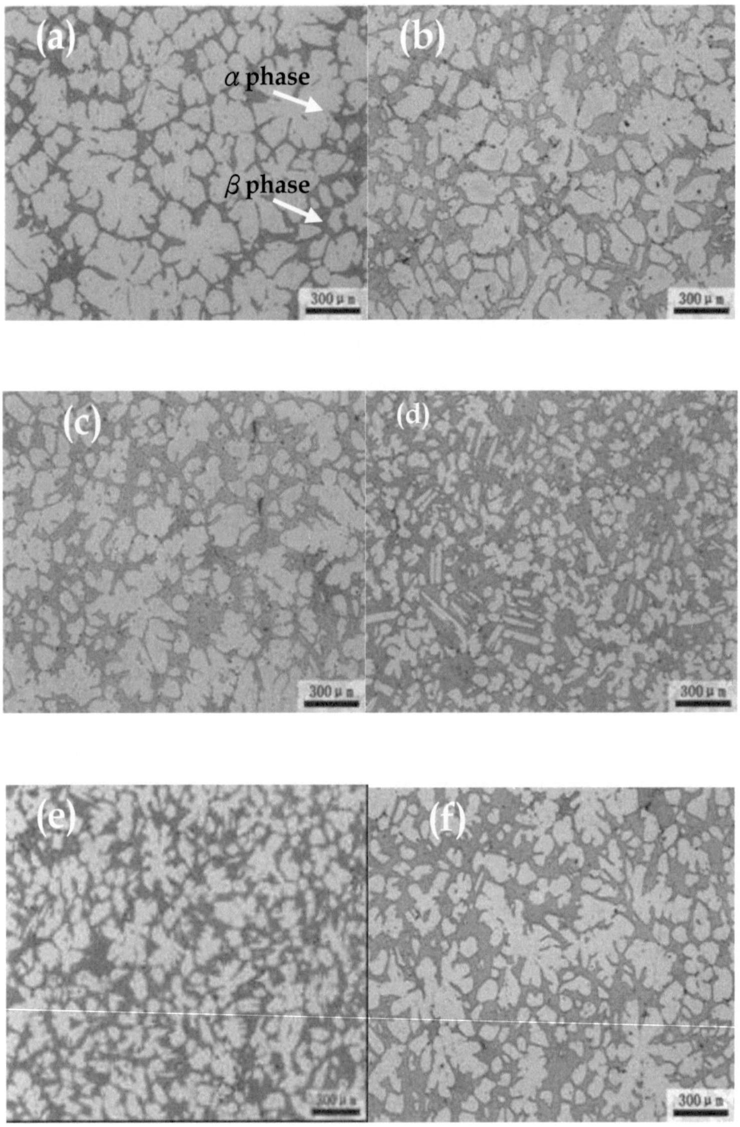


Fig. 13. Microstructures of specimens obtained with different ultrasonic vibration powers:(a) 0W (b) 50W (c) 110W (d) 170W (e) 210W (f) 260W.

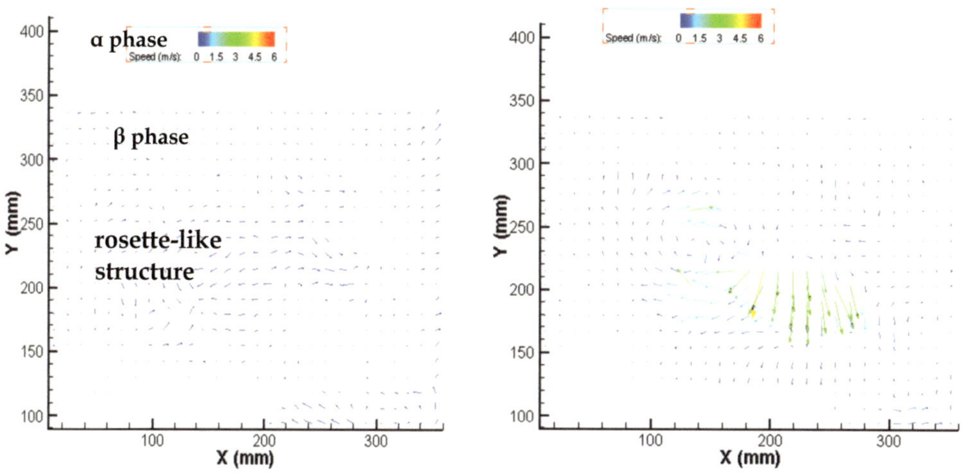


(a) 50W (b) 350W

Fig. 14. Effects of ultrasonic vibration power on fluid field by PIV physical simulation

## 5. Grain refinement of magnesium alloys

Magnesium alloys have extensive applications due to their comprehensive properties, such as low density, high specific strength, improved damping property and their recyclability. However, magnesium has bad plastic processing ability because of their HCP structure[33]. For magnesium alloys grain refinement is important as a fine grain size generally lead to improved mechanical properties and a more uniform distribution of secondary phases and solute elements on a fine scale which results in better machinability, good source finish, and excellent resistance to hot tearing and superior extrudability.

In the last few decades, the grain refinement of Magnesium alloys has been a particularly active topic and deserves more and more attention. A variety of methods have been developed to refine the magnesium alloys, such as rapid quenching, particle incubation, adding solute elements, imposing external fields and mechanical stirring. Among these methods, adding grain refiner (elements, master alloy) is known to be more effective for reducing the grain size of Mg-based alloys and have great importance on the industrial applications. Depending on whether they are alloyed with aluminum, magnesium alloys can be generally classified into two broad groups: aluminum free and aluminum bearing. Magnesium alloys containing zirconium or grain refined by zirconium such as ZE41, ZK60, WE43 and ML10. These are an important high value added class of alloys are based on the exceptional grain refining ability of Zirconium when added to aluminum free magnesium alloys. Because aluminum and zirconium form stable intermetallic phases, which are ineffective as nucleants for magnesium grains, the exceptional grain refining ability of zirconium does not occur in the aluminum bearing magnesium alloys.

Due to the importance of grain refinement to a broad range of aluminum and magnesium alloys, considerable work has been carried out for over half a century to determine the mechanisms by which grain refinement occurs. It is now generally accepted that both the potency of the nucleant particles (defined here as the undercooling required for nucleation,

$\Delta T_n$) and the segregating power of the solute ( defined as the growth restriction factor, $Q$) are critical in determining the final grain size. Easton and StJohn[34-35] developed a model that takes into account both $\Delta T_n$ and $Q$, and good agreement was found between this model and experimental results and proposed a semiempirical equation below for grain formation under small undercoolings:

$$d = \frac{1}{\sqrt[3]{fN_v}} + \frac{b' \cdot \Delta T_n}{Q} \tag{4}$$

Where $N_v$ is the number of relatively potent nucleant particles present in the melt and $f$ is the fraction of those particles that actually nucleate a grain.

Theoretically, Q was originally derived to be inversely proportional to the growth rate of the primary phase. More recently, it has been defined as the initial rate of development of constitutional undercooling with respect to fraction solid and can be estimated using the sum (GRF) of $mc_0(k-1)$ of the individual elements present in most wrought alloy systems, where $m$ is the slope of the liquids, $c_0$ is the concentration of the element, and $k$ is the partition coefficient. The higher GRF and solute element content is, the more obvious the effect of refining the solute elements in the alloy.

According to the equation (4), the addition of potent nucleant particles can lead to grain refinement of magnesium alloys. There is a necessary condition for the nucleant particles to act as heterogeneous nuclei, that is, the disregistry between low indexes planes of adjoining phases must be less than 15%. According to the disregistry model of two-dimensional lattices proposed by Bramfitt[36], the formula is：

$$\delta_{(hkl)_n}^{(hkl)_s} = \sum_{i=1}^{3} \{ | d[uvw]_s^i \cos\theta - d[uvw]_n^i | / d[uvw]_n^i \} / 3 \times 100\% \tag{5}$$

Where (hkl)s and (hkl)n are the low index planes of the matrix and nucleus, respectively,[uvw]s and [uvw]n are the low index orientations in (hkl)s and (hkl)n ,respectively, d[uvw]s and d[uvw]n are atomic spacing distances along [uvw],and θ is the angle between [uvw]s and [uvw]n. Some compounds, such as AlN, SiC, FeAl and AlMn, have also been to reported to be potential heterogeneous nucleation agents for Mg alloys, and AlN appears to be the most promising of all these compounds, because AlN has a simple HCP structure with lattice parameters of a=0.3111nm and c=0.4979nm, which are very close to the lattice parameters of the Mg matrix (a=0.3209nm, c=0.5211nm). The disregistry between AlN and a-Mg in the low index planes (0001) is 3.05%, respectively. Fig.15[37] show the microstructure of AZ31 alloys with AlN additions.

It is well known that Al-5Ti-B master alloy is an effective grain refiner in aluminum alloys. From the literature, GRF values of Ti, is far greater than the other alloying elements (e.g. Zr, Sr, Ca, etc.) values, and according to the equations(4), the crystal lattice mismatch between (0001) of TiB2 and (0001) of Mg is 5.6%(<9%) , the crystal face (0001) of TiB2 can be seen the heterogeneity nucleation basis of Mg phase .More and more researchers pay more attention on the effects of Al-Ti-B additions on the grain size of Mg-based alloys. Qi et al. [38] reports optimum average grain size and mechanical properties of AZ31 magnesium alloy with Al-5Ti-1B master alloy is obtained when the addition of Al-5Ti-1B master alloy is at 0.5wt% , shown in the Fig.16.

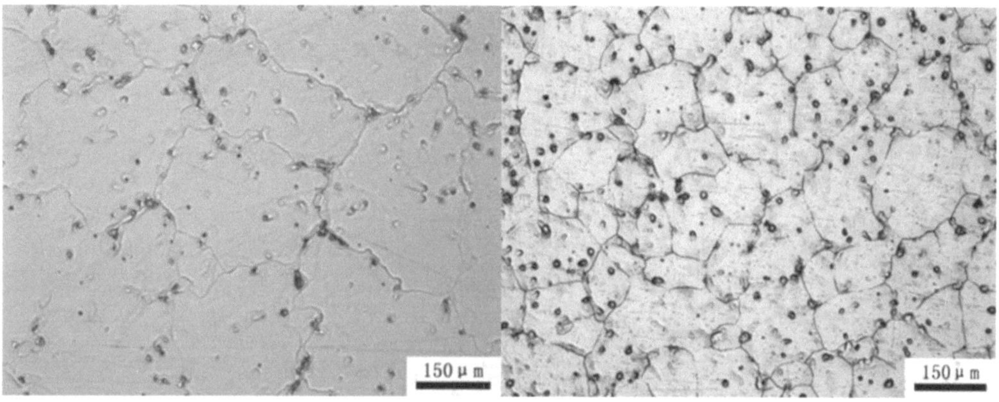


(a) without AlN addition; (b) 0.2wt% AlN addition

Fig. 15. Microstructures of AZ31 alloys

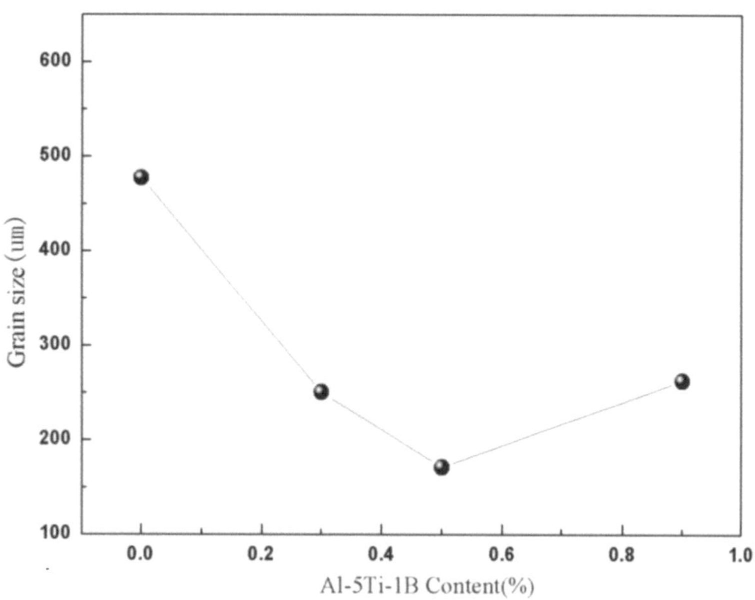


Fig. 16. The relationship between the content of Al-5Ti-1B master alloys and grain size of AZ31 alloy

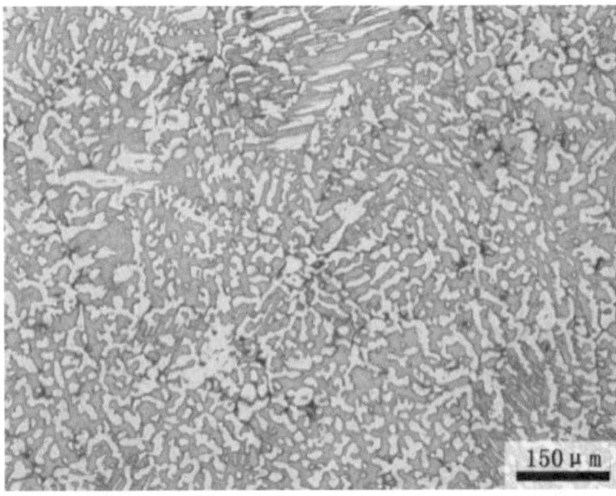


(a) 0V+0Ca

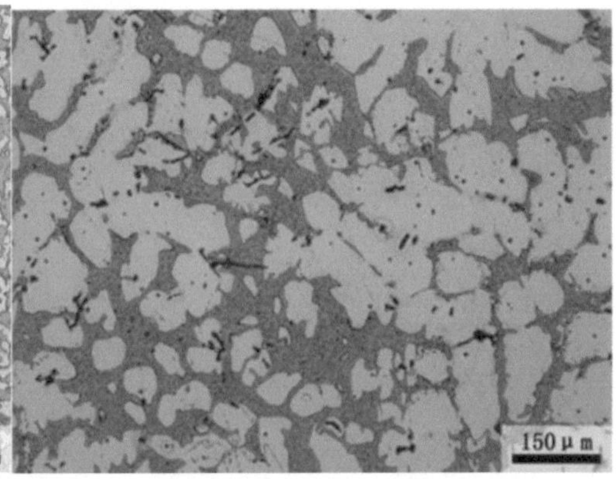


(b) 40V +0.5 Ca

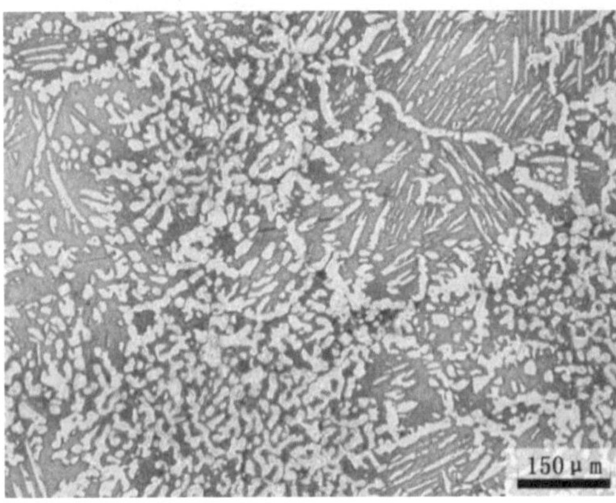


(c) 60V+0.5Ca

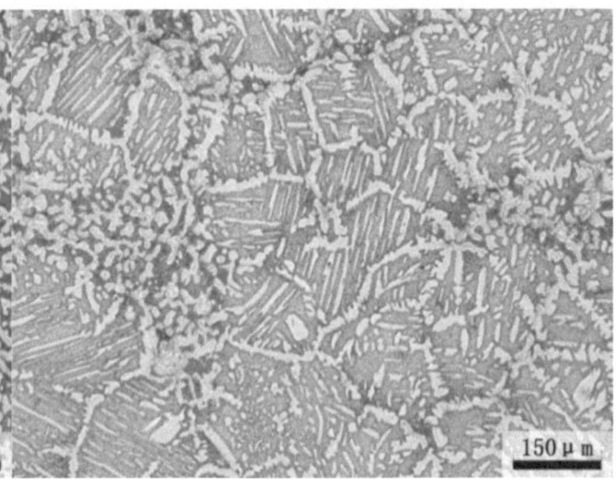


(d) 80V +0.5 Ca

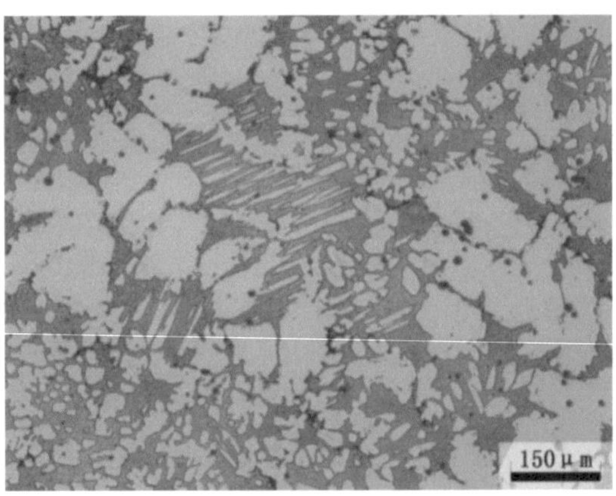


(e) 100V+0.5Ca

Fig. 17. Effects of Ca and electromagnetic stirring on the microstructure of Mg-8Li-3Al alloys

In recent years, the research about compound effects of alloying and external fields on grain refinement of magnesium attracts more and more attention. It is well known that the most important characteristic of magnetic field is its capacity to inject thermal and mechanical energy into materials without contact between the materials and the power source, which can produce driving, stirring, purifying or transmitting, leading to reduce the grain size and improve the mechanical properties. Hao et al.[39] have studied the couple effects of Ca and electromagnetic field on microstructure and mechanical properties of Mg-Li-Al alloys. In his work, when the electromagnetic stirring voltage is 80V, 0.5% Ca addition could make the microstructure fine and uniform (Fig.17) and the tensile strength was increased to 203.8Mpa.

## 6. The future of cast technology and quality improvement of magnesium alloys

Magnesium alloys have been called "the 21th century's engineering materials" for their high specific strength, high stiffness ratio, good machinability, good thermal conductivity and especially for their damping capability. However, the low mechanical properties and poor chemical properties, such as corrosion and creep resistance have restricted their extensive application. Despite these problems, the potential benefit of magnesium alloys has lead to a recent increase in demand for cast and wrought magnesium products. With this increase, the casting process is receiving significantly more attention from the standpoint of process optimization. Typical defects that occur during DC casting include both hot tearing as well as cold cracking that lead to downstream defects during subsequent processing operations, and are major sources which restrict the productivity of the process and its viability of alloys and ingot size. Modeling the stress-strain behavior and hot tearing of an AZ31 billet is able to quantitatively describe the evolution of temperature in the billet and quantitatively predict the development of residual stresses/strains. The application of EPM on the magnesium alloys can refine grain size and improve the material performance and this technology has become a helpful means to obtain high quality metal products. Ultrasonic treatment is one of the effective ways to improve the solidification structure of magnesium alloy and can improve the corrosion resistance and mechanical properties. A fine grain size generally leads to improved mechanical properties and structural uniformity of magnesium alloy, and more attention should be paid for the mechanism of grain refinement of magnesium alloys.

## 7. Acknowledgements

Financial support from the Program of New Century Excellent Talents of the Ministry of Education of China (NCET-08-0080), the National High Technology Research and Development Program ("863"Program) of China (2009AA03Z525), the Science and Technology Fund of Dalian City (2009J21DW003), and the Fundamental Research Funds for the Central Universities (DUT11ZD115) are gratefully acknowledged.

## 8. References

[1] P.T. McGlade, 6th Aus. Asian Pacific Conf. On aluminum cast house technology, sydeney, Australia, 1999,P.R.Whiteley and J.F. Grandfield;

[2] D.C. Weckman, P. Niessen. A numerical simulation of the D.C. continuous casting process including nucleate boiling heat transfer[J]. Metallurgical and materials transactions B. 1982,13:593-602;

[3] D. Adenis, K. Coats, D.RAGONE. An analysis of direct-casting process by numerical methods[J]. Journal of the insttitute fo metals. 1963,91:395-403;
[4] S.G. Hibbins. In light metals 1988 metaux legers, M.Sahoo and C.C. FRADET, eds., TMS-CIM, Calgary, AB, Canada, 1988:265-280;
[5] H. Fossheim, E.E Madsen. Light metals, 1979, Newe Orleans, La, 1979, W.S. Peterson, ed. TMS-AIME, Warrendale, PA,1979:695-720;
[6] H.G. Fjaer, A. Mo. Metall. Alspen-A mathematical- model for thermal-stresses in direct chill casting of aluminum billets. Metallurgical transactions B. 1990, 21: 1049-1061;
[7] Y. Watanabe, N. Hayashi. Light metals 1996, Anaheim, CA, 1996, W.Hale,ed., TMS, Warrendale, PA:979-984;
[8] J. Sengupta, S.L. Cockcroft, D. Maijer, M. Wells, A. Larouche. J. Light Met., 2002,2:137-148;
[9] J. Senqupta, S. Cockcroft, D. Maijer, M. Wells, A. Larouche. On the development of a three-dimensional transient thermal model to predict ingot cooling behavior during the start-up phase of the direct chill-casting process for an AA5182 aluminum alloy ingot[J]. Metallurgical and materials transcations B. 2004,35B:523-540;
[10] J.M. Drezet, M.Rappaz. Light metals 2001, New Orleans, La,2001, J.L. Anjier, ed., TMS. Warrendale, PA,2001:887-894;
[11] M. Rappaz, J.M. Drezet, M. Gremaud. A new hot-tearing criterion[J]. Metallurgical and materials transactions A. 1999, 30A: 449- 455;
[12] A. Stacey, H.J. MacGillivary, G.A. Webster, P.J. Webster, K.R.A. Ziebeck. Measurement of residual stresses by neutro diffraction[J]. Journal of strain Analysis, 1985,20:93-100;
[13] P.J. Webster, X.D. Wang, W.P. Kang, G. Mills. Experimental verification of residual stress models using neutron strain scanning[C]. Campbell J,editor. Modelling of casting, welding and advanced solidification process Ⅶ. London, UK: The minerals, Metals & Materials Society;1995:311-318;
[14] M.J. Schmank, A.D. Krawitz. Measurement of a stress gradient through the bulk of an aluminum alloy using neutrons[J]. Metallurgical Transactions A. 1982, 13:1069-1076;
[15] J. Smith, M.N. Bassim, C.D. Liu, T,M. Holder. Measurement of crack tip strains using neutron diffraction[J]. Engineering Fracture mechanics. 1995,52:843-851;
[16] H. Hao, D. Maijer, R. Rogge. Investigation of residual strains by neutron diffraction in an AZ31 direct chill cast billet[J]. NDT&E International. 2009,42:704-712;
[17] M.M. Hamdi, A. Mo, H.G. Fjaer. Metall. Tearsim: A two-phase model addressing hot tearing formation during aluminum direct chill casting[J]. Metallurgical and materials transacations A. 2006,37A:3069-3083;
[18] D.J. Lahaie, M. Bouchard. Metall. Physical modeling of the deformation mechanisms of semisolid bodies and a mechanical criterion for hot tearing[J]. Metallurgical and materials transactions B. 2001,32:697-705;

[19] H. Hao, D.M. Maijer, M.A. Wells, A. Phillion, S.L. Cockcroft. Modeling the stress-strain behavior and hot tearing during direct chill casting of an AZ31 magnesium billet. Metallurgical and materials transactions A, 2010,41A:2067-2077.
[20] M. Yoshida. Electromagnetic casting update to date and future[J]. ISIJ, 1987,73:403-410;
[21] Z.N. Getselev. Electromagnetic ingot mold continuous casting[J]. Journal of metals. 1971, 23:38;
[22] S. Asai, T. Kozuka, I. Muchi. Process development and stability analysis of horizontal electromagnetic casting method[J]. Iron and Steel. 1986,16:68-74;
[23] Z.Q. Cao, F. J, X.G Zhang, H. Hao, J.Z. J. Microstructures and mechanical characteristics of electromagnetic casting and direct-chill casting 2024 aluminum alloys[J]. Materials science and engineering. 2002,A327:133-137;
[24] L. Pang, X.G. Zhang, Z. Ren, L. Sui, T. Zhang, H. Hao. Effects of medium-frequency electromagnetic field on solidification process of semi-continuous casting AZ31 magnesium alloy[J]. Materials research innovations. 2009,13:107-111;
[25] Z. Ren, X.G Zhang, H. Hao, L. Sui, Y.T. Ma, J.Z Jin. Effect of middle-frequency electromagnetic field on the semi-continuous casting for ZK60 billets[J]. Journal of wuhan university of technology. 2010,25:449-453.
[26] O.V. Abramov. Ultrasonic treatment of light alloy melts[M]. CRC Press, Boca Raton, FL,1994:289-329
[27] A. Ramirez, M. Qian, B. Davis, T. Wilks, D.H. Stjohn. Potency of high-intensity ultrasonic treatment for grain refinement of magnesium alloys[J]. Scripta Materialia. 2008,59:19-22;
[28] G.I. Eskin. Ultrasonic treatment of light alloy melts, Gordon& Breach, Amsterdam, 1998:1-25,135-240;
[29] J.D. Cutnell, K.W. Johnson. J.Appl.Phys.1966,37:254-257;
[30] M.C. Flemings, Solidification Processing , McGraw-Hill, New York, 1974:300;
[31] M Q, A. Ramirez, A. Das. Ultrasonic refinement of magnesium by cavitation: Clarifying the role of wall crystals[J]. Journal of crystal growth. 2009,311:3708-3715;
[32] L. Yao. Research on microalloying and solidification control with external field of Mg-Li series alloys[D].Dalin, Dalian university of technology. 2010.(In chinese)
[33] B.L. Mordike, T. Ebert. Magnesium properties-applications-potential[J]. Materials science and engineering. 2001,A302:37-45;
[34] M.A. Easton, D.H. Stjohn. A model of grain refinement incorporating alloy constitution and potency of heterogeneous nucleant particles[J]. Acta mater. 2001,49:1867-1878;
[35] M. Easton, D. Stjohn. An analysis of the relationship between grain size, solute content, and the potency and number density of nucleant particles[J]. Metallurgical and materials transactions A. 2005,36A:1911-1920;
[36] B.L. Bramfitt. The effect of carbide and nitride additions on the heterogeneous nucleation behavior of liquid iron[J]. Metallurgical and materials transaction B. 1970,1:1987-1995;

[37] S.W. Gu, H. Hao, C.F. Fang, S.H. Ji, X.G. Zhang. Effects of AlN particles and electromagnetic stirring on as-cast structure of AZ31 alloys[J]. Materials science forum. 2011, 675-677:771-774;

[38] G.H. Qi, C.F. Fang, H. Hao, X.G. Zhang. Effects of Al-5Ti-1B master alloy on microstructure and properties of AZ31 magnesium alloy[J]. Special casting and nonferrous alloys.2010,30:266-269;(In Chinese)

[39] H. Hao, L. Yao, S.W. Gu, L.J. Ma, X.G. Zhang. Effects of Ca and electromagnetic stirring on microstructures and mechanical properties of Mg-Li-Al alloys[J]. Transactions of nonferrous metals society of china. 2010,20:388-392;(In Chinese)

# 2

# Surface Modification of Mg Alloys AZ31 and ZK60-1Y by High Current Pulsed Electron Beam

Gao Bo, Hao Yi, Zhang Wenfeng and Tu Ganfeng
*School of materials and metallurgy, Northeastern University, Shenyang*
*China*

## 1. Introduction

The quality of Mg alloys with high specific modulus and specific strength is the lightest in the structural materials. Their density is about 2/3 of aluminium alloys and 1/4 of steels. The weight of whole structural materials is decreased drastically owing to some components or parts produced by Mg alloys. Thus, Mg alloys are widely used in aerospace, weapons, automobile and other fields [1-3]. Meanwhile, Mg alloys has many other advantages, such as excellent electromagnetic shielding performance, shock absorption ability, electric and heat conductivity, etc [4-6]. However, the chemical stability of Mg is very low, and its electrode potential is negative (-2.34V). As a result, the corrosion resistance of Mg alloys is poor in acid and neutral mediums. Furthermore, other properties of Mg alloys, wear resistance, hardness and resistance to high temperature, are also poor. Consequently, the superiority of Mg alloys in the application is restricted to some extent. Nowadays, the researches are concentrated on the improvement of hardness, wear and corrosion resistance of Mg alloys. Energy beam surface modification is an important developing direction, such as plasma micro-arc oxidation [7-11], laser surface treatment [12-16], ion beam surface modification [17-19], etc.

A.V. Apelfeld et al [10] have studied oxide protective coatings on the surface of Mg alloys obtained by micro-arc oxidation (MAO). A model of micro-arc coating formation is proposed. For Mg alloys, the structure of MAO coating plays an important role in improvement of corrosion resistance.

The research team (Y.M. Wang et al [11]) has investigated that dense oxide coatings formed in alkaline silicate electrolyte with and without titania sol addition are fabricated on AZ91D alloy using micro-arc oxidation. It reveals that the coating thickness decreases from 22μm to 18 μm with increasing concentration of titania sol from 0 to 10 vol. %. Electrochemical tests show that the Ecorr of Mg substrate positively shifts about 300-500 mV and Icorr lowers more than 100 times after micro-arc oxidation.

The literature [15] (A.K. Mondal et al) has reported that Mg alloy ACM720 is subjected to laser surface treatment using Nd:YAG laser in argon atmosphere. This treatment is beneficial for enhancing the corrosion and wear resistance of the alloy. The improved corrosion resistance is attributed to the absence of second phase $Al_2Ca$ at the rain boundaries, microstructural refinement and extended solid solubility, particularly of Al, in α(Mg) matrix owing to rapid solidification. The laser treatment also increases surface hardness two times and reduces the wear rate considerably due to grain refinement.

J. Dutta Majumdar and I. Manna [16] have researched that the mechanical properties of laser-surface-alloyed AZ91, a magnesium-based alloy (Mg-9Al-0.9Zn) with nickel. Laser surface alloying is carried out using a continuous wave $CO_2$ laser. As a result, laser surface alloying leads to the formation of a dispersion of intermetallics of Mg and Ni ($MgNi_2$) in an Mg matrix with an improved Young's modulus (45-85 GPa, as compared to the 45 GPa of as-received substrate) and improved wear resistance.
J.X. Yang et al [19] have found that thin carbon nitride (CN) coating can be deposited on Mg alloy substrate by ion beam assisted deposition. Through adjusting parameter, the coating with high N/C ratio of 0.38 is produced. The CN coating is a composition of amorphous $C_3N_4$ and improves the roughness of naked substrate from 32.9 nm to 28.9 nm. Moreover, the percentage increases of hardness and elastic modulus induced by coating are more than 90.6% and 82.8%, respectively.
Surface treatment of electron beam on metallic materials, such as steels, pure Al and Al alloys, NiTi alloy, has recently been investigated by domestic and foreign researchers [20-26]. It is found that metastable structures are formed on treated material surface and the hardness as well as wear and corrosion resistance is enhanced after electron beam treatment. However, the investigation on Mg alloys after electron beam treatment is reported rarely in literatures. In our group, the change in microstructure and properties of Mg alloys (AZ31, ZK60-1Y) has been researched in detail before and after HCPEB treatment.

## 2. Raw materials, devices and detection methods for experiment

### 2.1 Raw materials for experiment

The research objects of this passage are magnesium alloys AZ31 and ZK60-1Y. The chemical compositions of two Mg alloys refer to the table 2.1.

| Elements | Al | Zn | Zr | Y | Mn | Si | Cu | Fe | Ni | Mg |
|---|---|---|---|---|---|---|---|---|---|---|
| AZ31 | 2.5~3.5 | 0.6~1.4 | — | — | 0.2~1.0 | 0.1 | 0.05 | 0.005 | 0.005 | balance |
| ZK60-1Y | — | 6 | 0.5 | 1 | — | — | — | — | — | balance |

Table 2.1. Chemical compositions of Mg alloys AZ31and ZK60-1Y (*wt.%*)

### 2.2 The structure and technological parameters of the high current pulsed electron beam (HCPEB) device

The Fig. 2.1 refers to "Nadezhda-2" type HCPEB device [27] made in Russia. It consists of four parts, such as electron gun, vacuum system, power supply control system and diagnostic system. Vacuum system includes vacuum pump set (a set of molecular pump & a set of mechanical pump), vacuum chamber, vacuum valve as well as pipes for inputting cooling water and waste gas emission. Power supply control system includes high-voltage pulse generator and magnet field triggering power supply, etc. Diagnostic system includes relevant instruments and meters, such as operating vacuum measurement, electron beam, cathode accelerating voltage and average energy density, etc. Electron gun includes cathode, anode, spark source, rogowski coil and magnetic coil, etc. The electron gun used for generating high current pulsed electron beam is the core part of the whole equipment.

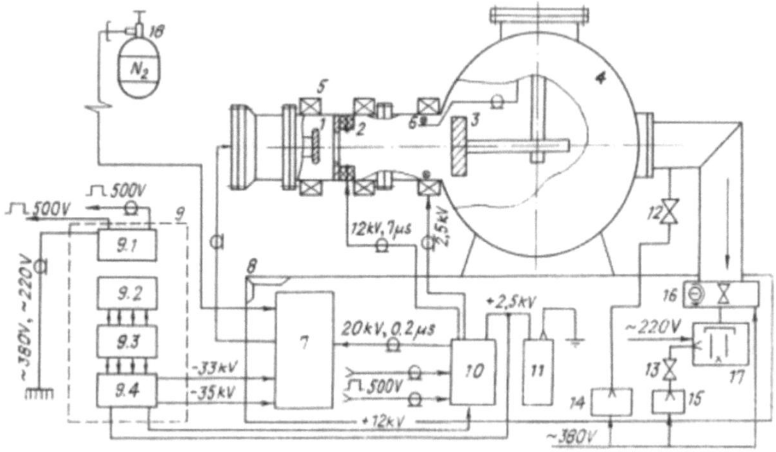


1-cathod; 2-spark source; 3-collector; 4-vacuum chamber; 5-solenoid; 6-rokovsky coil; 7-pulsed high-voltage generator; 8-bracket; 9-electricity controller; 10-pulses trigger; 11-capcitor; 12, 13-manual vacuum valve 1, 2; 14, 15-pump 1, 2; 16–electromagnetism valve, 17-molecular pump, 18-nitrogen.

Fig. 2.1. Schematic diagram of high current pulsed electron beam (HCPEB) system

"Nadezhda-2"type HCPEB device can generate a special form of electron beam. In this experiment, the main technological parameters are shown in table 2.2, in which the energy density of electron beam can be regulated through changing the capacitance of the high-voltage pulse generator, cathode accelerating voltage and adding magnetic field.

| Accelerating voltage /kV | Energy density /J/cm$^2$ | Pulse time /μs | Pulse frequency /Hz | Pulse number /Number | Target distance /mm | Beam spot diameter /mm |
|---|---|---|---|---|---|---|
| 27 | 3 | 1 | 0.1 | 5,10,15 | 140 | ϕ60 |

Table 2.2. Technological parameters for HCPEB equipment

### 2.3 Detection methods

#### 2.3.1 Analysis of microstructure

The morphologies of surface and sectional microstructures of Mg alloys were analyzed through scanning electron microscope(SM-5600LV, SSX-550). Phase transformation of the treated surface of Mg alloys was conducted by using X-ray diffractometer (XRD-6000) with step size of 0.02° and scanning range (2θ) of 20°~100°. The experiment adopts a Cu Kα radiation source.

#### 2.3.2 Friction and wear performance

The dry friction and wear tests of Mg alloy AZ31 surface were finished using a ball-on-flat apparatus at environmental temperature of 18~22°C. The sample surfaces were cleaned by acetone ultrasonic before friction and wear test. WC-Co balls with diameter 5mm were used as the sliding counterpart in all tests. The applied load was 5 N with a sliding velocity of 1 mm/s. The total sliding length and stroke length were 5mm and 1.2m, respectively.

Drying friction and wear testing of Mg alloy ZK60-1Y was conducted with a pin-on-disc type machine (MG-2000) at room temperature of 25°C. The samples used for experiment were cut into cylinders with diameter of Φ6mm and height of 12mm. The counterpart discs were made of stainless steel (1Cr18Ni9) with surface hardness of 192HV and surface roughness of 1μm ($R_a$). The applied load was 10N. The rotation speed was 250 r/min and the friction time was 10min.

#### 2.3.3 Corrosion resistance

Polarization curves were performed in the EG&G M273 system. The experiment adopted three-electrode system. The reference electrode was saturated calomel electrode (SCE), the auxiliary electrode was Pt electrode, and the samples were working electrode. The test was performed in 5% NaCl solution and the corrosion testing surface area was $1cm^2$.

## 3. Experiment results and discussion

### 3.1 Surface modification of Mg alloy AZ31 by HCPEB

#### 3.1.1 Surface morphology

Fig. 3.1 gives the surface SEM morphologies of Mg alloy AZ31 after HCPEB treatment with energy density of $3J/cm^2$. Fig. 3.1 (a) refers to the surface morphology of Mg alloy AZ31 after HCPEB treatment of 5 pulses. It can be apparently observed that a typical morphology emerges on the AZ31 Mg alloy surface after HCPEB treatment, namely wavy morphology after complete melting. With the pulse number increasing to 10 pulses, the "crater" morphology and few twins are found on the treated surface shown in the Fig. 3.1 (b). Fig. 3.1 (c) shows surface morphology of Mg alloy AZ31after 15 pulse treatment, it is found that the "crater" morphology disappears after repeated melting and a large number of twins are formed. The 15-pulsed sample surface tends to be smooth. The formation of twins is possibly induced by a lot of residual stresses on the surface after HCPEB treatment [28].

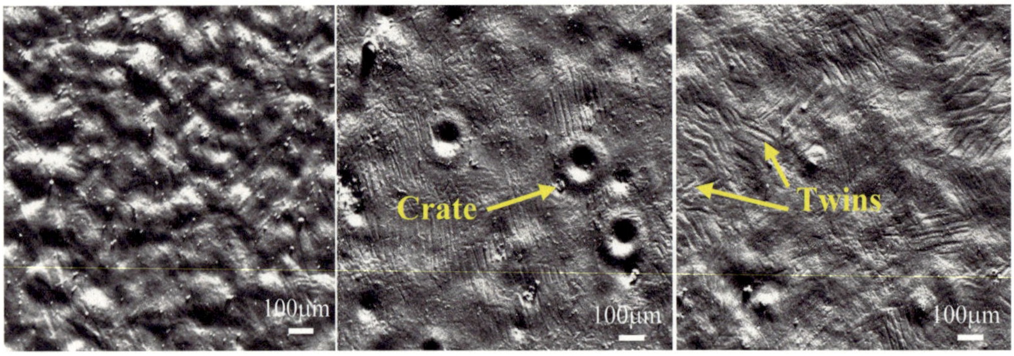


(a) 5 pulses (b) 10 pulses (c) 15 pulses

Fig. 3.1. Surface SEM morphologies of AZ31 Mg alloy after HCPEB irradiation with different pulses

#### 3.1.2 Phase analysis

Fig. 3.2 indicates XRD patterns of Mg alloy AZ31 before and after HCPEB treatment and local analysis of (11-20) peaks. From the figure, it is found that the diffraction peaks of Mg obviously move towards high-angle direction after 5 and 10 pulse treatments, as clearly shown in the enlarged figure. The moving of Mg diffraction peaks is more obvious with the increase of pulse number. This is generated by rapid heating and cooling process induced by HCPEB on the surface of Mg alloy. The substitutional solid solution of Al atoms replacing Mg atoms is formed in the Mg lattices, and the lattice constants are reduced as increasing solid solubility. To sum up, high-angle moving of Mg diffraction peaks is attributed to the formation of saturated solid solution of Mg [29].

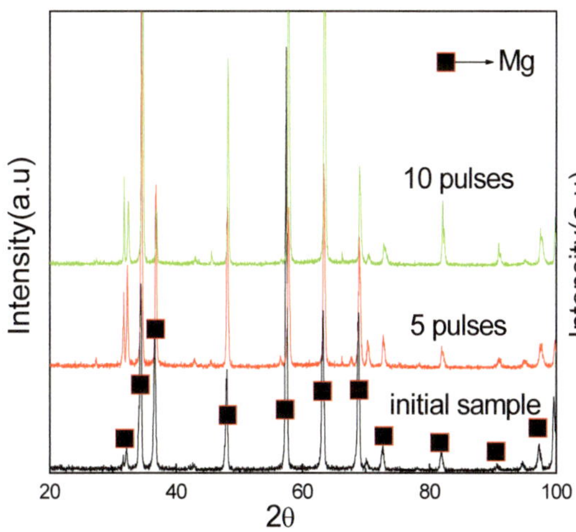


(a) XRD patterns

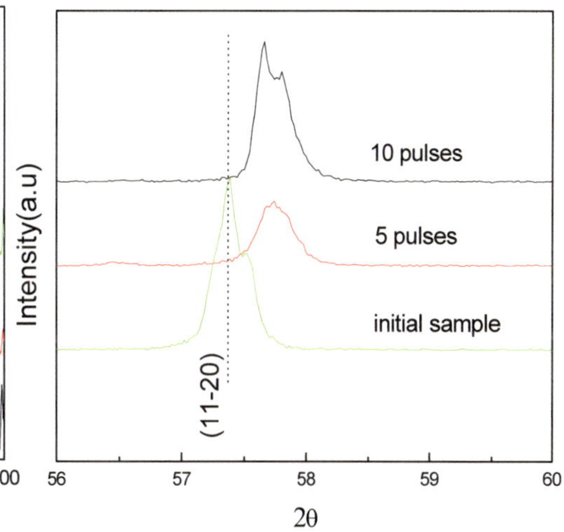


(b) Local analysis of (11-20) peaks

Fig. 3.2. XRD patterns of Mg alloy AZ31before and after HCPEB treatment

### 3.1.3 Friction and wear property of Mg alloy AZ31

Fig. 3.3 indicates the evolution of friction coefficients with friction time and change of average friction rate with the number of pulses for Mg alloy AZ31 before and after HCPEB treatment. It can be seen that the friction coefficients of initial samples are only about 0.15 at the beginning (as seen in Fig.3.3(a)). It is due to a layer of hard MgO film on the surface. The oxide film is worn away after 5 minutes and the wear enters into severe wear stage, thus, the friction coefficients suddenly rise to 0.32~0.37. But the friction coefficients of treated samples are relatively stable (between 0.25 and 0.27). On the one hand, the oxide film on the sample surface is damaged during HCPEB process, then MgO film can not rapidly formed in the vacuum. On the other hand, the higher roughness on the surface will also lead to the increase in the friction coefficients. Compared to initial samples, the friction coefficients of Mg alloy AZ31 are obviously reduced after HCPEB bombardment. Additionally, the wear rate is reduced by a factor of 6.7 after 15 pulse treatment (as seen in Fig.3.3 (b)), so the wear resistance of Mg alloy AZ31 is increased significantly. It is indicated that there is a large potential in the application of Mg alloy AZ31 after HCPEB treatment.

Fig. 3.4 gives SEM morphologies of wear grooves on the surface of Mg alloy AZ31before and after HCPEB treatment. Compared to wear groove morphology on the surface of initial

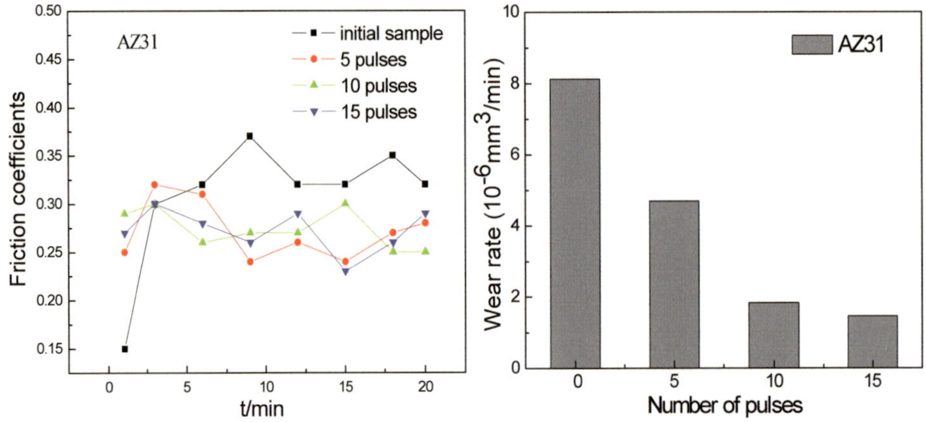


(a) Friction coefficients (b) Average friction rate

Fig. 3.3. Evolution of friction coefficients with friction time and change of average friction rate with the number of pulses for Mg alloy AZ31 before and after HCPEB treatment

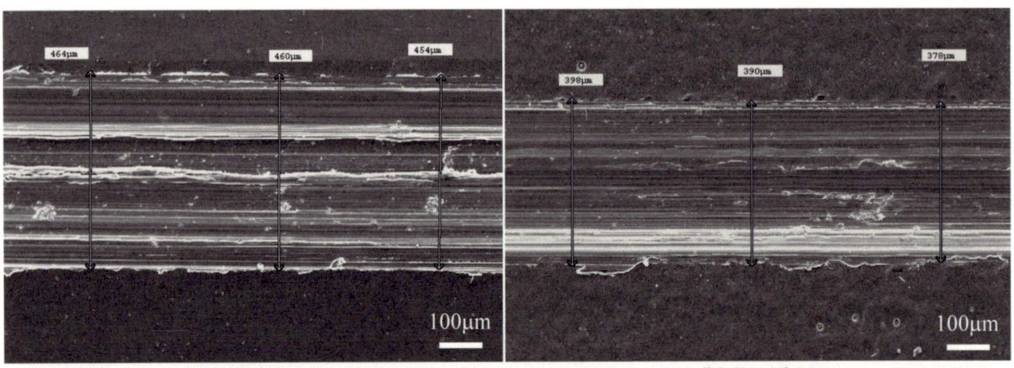


(a) initial sample (b) 5 pulses

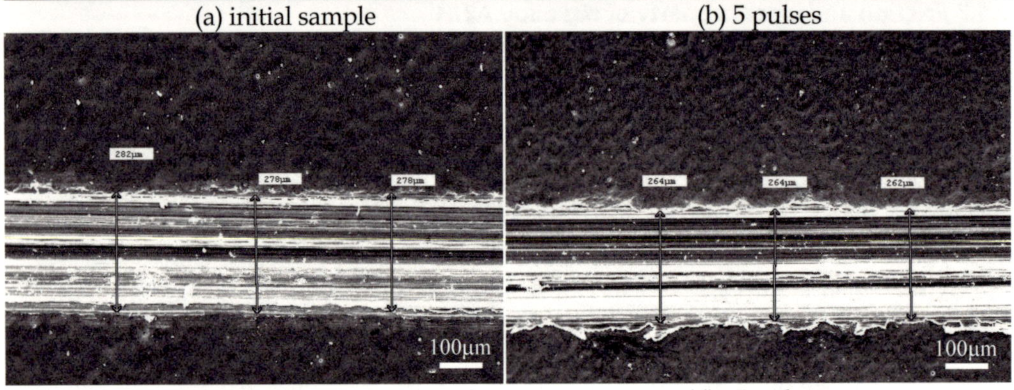


(c) 10 pulses (d) 15 pulses

Fig. 3.4. SEM morphologies of wear grooves for Mg alloy AZ31 before and after HCPEB treatment with energy density of ~3J/cm2

sample, the width of wear grooves on the surface of Mg alloy AZ31 is decreased obviously after HCPEB treatment, which is reduced from 460μm of initial sample to 280μm of 10-pulse treated sample. The wear form is the typical abrasive wear for Mg alloy AZ31. The hardness of matrix is far smaller than that of WC grinding ball, so the cutting form is focused on the wear surface and the groove morphology is formed. After HCPEB treatment, the residual compressive stress is left on the surface, leading to the increase in hardness of HCPEB-treated Mg alloy. As a result, the wear rate and wear volume are reduced, and the width of wear grooves is also decreased. Meanwhile, the main composition of wear debris is the oxides of Mg and Al through EDS analysis of SEM. It is caused by oxidization resulting from the fact that the wear test is performed in the atmosphere, and a lot of heat is accumulated on the sample surface during the severe friction process.

### 3.1.4 Corrosion resistance of Mg alloy AZ31

With focus on the problem of poor corrosion resistance of AZ31 Mg alloy in practical application (contact with atmosphere or sea water), this paper applies HCPEB to perform surface treatment in order to analyze the chemical composition of the surface remelted layer and discuss the change in corrosion resistance of Mg alloy AZ31 surface in 5%NaCl solution.

#### 3.1.4.1 Surface composition analysis of Mg alloy AZ31 before and after HCPEB surface treatment

The chemical composition of alloy surface plays an important role in corrosion process. The changes of Mg, Zn and Al elements are analyzed by SEM. It is found that the variation of Zn is very small after HCPEB treatment. The changes of Mg and Al are mainly given in Fig. 3.5. It can be seen that the content of Mg on the surface is reduced after HCPEB treatment and reaches maximum reduction after 5 pulse treatment, which is due to Mg evaporation in the molten state. As the number of pulse increases to 10 pulses, the melted layer is thickened. The Mg contained underneath new melting liquid will spread or evaporate to the upper layer, so the Mg content on the surface of Mg alloy AZ31 will rise again. With further increase in pulse number, the Mg at the surface layer loses balance as well as the Mg inside the equipment, and the Mg at the surface layer evaporates, so the content of Mg is reduced again. But, the change of Al content is on the contrary to that of Mg. The increased Al content at the surface layer will be undoubtedly favorable to the formation of compact oxide film and the enhancement of corrosion resistance on the alloy surface. Therefore, in order to acquire the surface that contains high content of Al, it is crucial to choose proper pulse number and energy density.

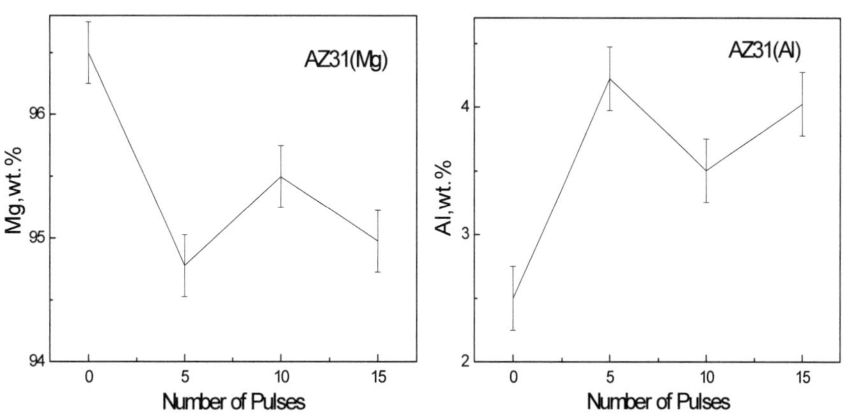


Fig. 3.5. Content of Mg and Al on the surface of Mg alloy AZ31 after HCPEB irradiation

#### 3.1.4.2 Electrochemistry testing for Mg alloy AZ31

Fig. 3.6 shows the potentiodynamic polarization curves of Mg alloy AZ31 before and after HCPEB treatment. From the measuring result, we can see that the Ecorr of modified sample is moved to -1360mV. Through Tafel straight line fitting of cathode and anode, it is discovered that the corrosion current is reduced but the polarization resistance is increased. The reason is that HCPEB treatment leads to rich Al on the AZ31 Mg alloy surface and the formation of compact oxide film. Compared to the initial sample, the corrosion resistance of HCPEB-treated sample is increased. It can be also discovered that the overpotential (the difference between applied potential and Ecorr) reaches about 300mV, namely breakthrough occurs. For one thing, the surface protective film is comparatively thin. For another thing, the applied etching solution is 5% NaCl, in which chloride ion has strong penetrability. Hence the protective film is punctured rapidly. It also illustrates that HCPEB treatment can not be solely applied as the final process of the sample treatment, but it should be applied practically together with other processes.

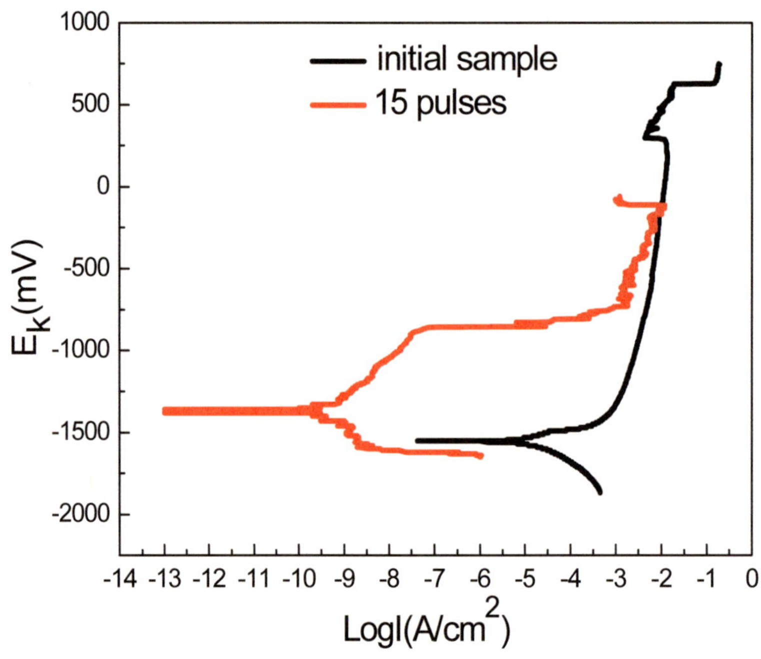


Fig. 3.6. Potentiodynamic polarization curves of Mg alloy AZ31 before and after HCPEB treatment (15 pulses)

**Analysis of corrosion mechanism**

The corrosion of Mg alloy AZ31 is mainly divided into two stages: 1) the initial stage is controlled by galvanic corrosion; 2) the expansion stage is controlled by pitting.
The initial stage of corrosion:
In the neutral environment, a layer of protective $Mg(OH)_2$ thin film is formed on the surfaces of pure Mg and Mg alloy at the initial stage of corrosion, as shown in the following equation.

$$Mg + 2H_2O \rightarrow Mg(OH)_2 + H_2 \quad (1)$$

Although Mg $(OH)_2$ has better protectiveness in the atmospheric environment, it alters [30] easily with the changes of electrochemistry and environmental conditions. The oxide grows initially from the crystal boundaries and extends into crystal grains. However, as for Mg alloy, the potential of crystal boundaries is higher than that of crystal grains, so the crystal boundaries form a galvanic couple with the crystal grains. The corrosion starts from the crystal boundaries and extends into crystal grains. Because the proportion of anode and cathode is larger and larger, the corrosion speed is accelerated with further corrosion. The rapid cooling (~$10^8$K/s) induced by HCPEB makes crystal grain fine and reduces the proportion of anode and cathode. Therefore, the corrosion current of the modified sample is smaller than that of the initial sample. Meanwhile, the increase in the solid solubility of Al will also increase the integral polarization resistance of the material. The first stage is finished when the material surface is covered with a layer of compact and stable magnesium hydroxide film.
The expansion stage of corrosion:

$$Mg^{2+} + 2Cl^- \rightarrow MgCl_2 + 2e^- \quad (2)$$

In the environment where the chloride ion exists, a sort of soluble magnesium salt forms in the boundary of metal and solution, as shown in the equation (2). This soluble magnesium salt ruins the protective film of magnesium hydroxide, and then the exposed metal generates self-solvent reaction with electrolyte in the damaged film and the corrosion speed is accelerated thereupon. The partial attack causes the pitting formed on the alloy surface. Once the pitting starts, it will expand at a very fast speed. Thinking over from the angle of kinetics, the pitting formation rate of the treated sample is smaller than that of the initial sample. It should be contributed to stability and compactness of oxide film formed on the surface of modified sample. It is equivalent that the anode dissolution speed is slowed down, thus Icorr is reduced remarkably.

### 3.2 Surface modification of Mg alloy ZK60-1Y by HCPEB

As for surface modification of Mg alloy ZK60-1Y by HCPEB treatment, the experiment process and parameters are similar to those of Mg alloy AZ31, namely accelerating voltage 27kV, energy density 3J/cm$^2$.

#### 3.2.1 Surface morphology

Fig. 3.7 refers to surface SEM morphologies of Mg alloy ZK60-1Y before and after HCPEB treatment. Before electron beam treatment, the original structure of Mg alloy ZK60-1Y is composed by eutectic structure and matrix $\alpha$(Mg), and the eutectic structure shows discontinuous distribution (white contrast in the figure), as shown in the Fig. 3.7 (a). After HCPEB treatment, the crystal boundaries on the surface of Mg alloy ZK60-1Y become less obvious, and the composition tends to be uniform. It is because that the HCPEB bombardment results in melting of the alloy surface, and it is too late to be redistributed and solidified for the elements owing to subsequent rapid solidification. At the same time, the typical "crater" morphology is formed on the surface of Mg alloy (Fig. 3.7 (b)). It is due to the ZK60-1Y Mg alloy surface is under the action of pulsed beam current, the subsurface is melted first and liquid is erupted outwards from inside, which is similar to the volcanic eruption. It is one reason for the formation of "crater" morphology [31]. With increasing pulse number, more energy is absorbed by the surface of Mg alloy, as a result, the "crater" quantity is reduced and the surface becomes much smoother, as shown in the Fig. 3.7 (c-d).

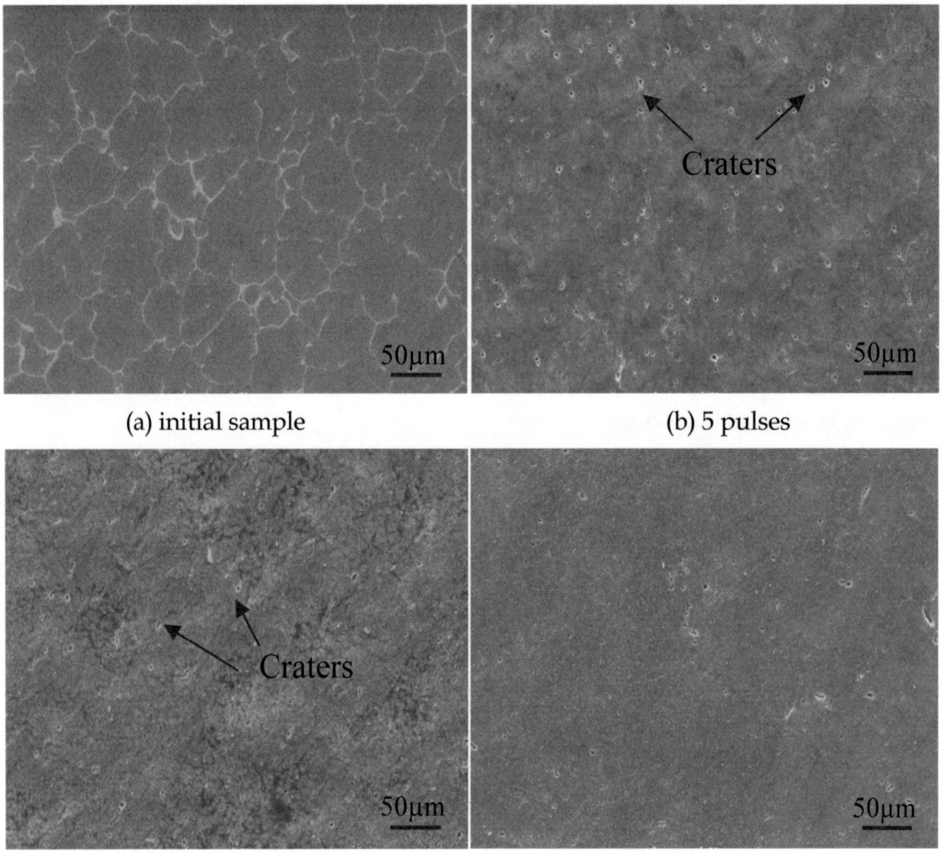


(a) initial sample (b) 5 pulses

(c) 10 pulses (d) 15 pulses

Fig. 3.7. Surface SEM morphologies of Mg alloy ZK60-1Y before and after HCPEB treatment

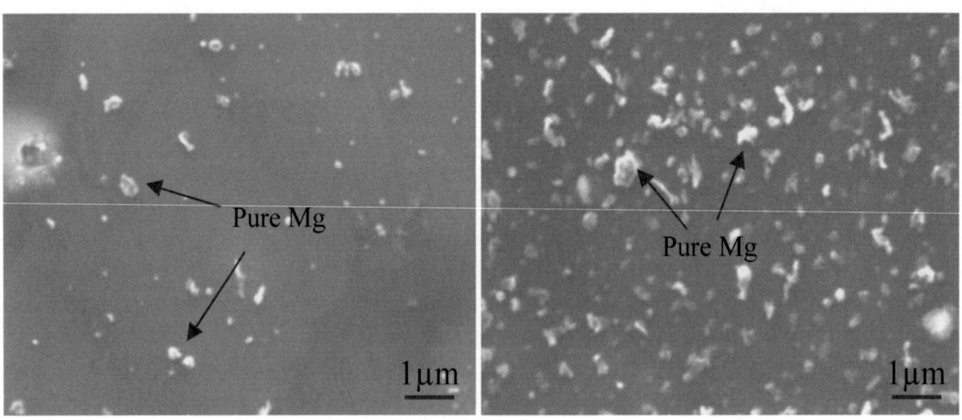


(a) 5 pulses (b) 15 pulses

Fig. 3.8. Pure Mg particles on the surface of Mg alloy ZK60-1Y after HCPEB treatment with accelerating voltage of 27kV

Fig. 3.8 shows pure Mg particles on the surface of Mg alloy ZK60-1Y after HCPEB treatment. From the figure, it can be seen that the pure Mg particles (the result of EDS analysis) are distributed on the surface of Mg alloy after HCPEB treatment and more pure Mg particles emerge on the alloy surface after 15 pulse treatment. The reason can be explained as follows: Mg belongs to the low melting point metal and its vapor pressure is also very low compared with Zr and Y elements. Under HCPEB bombardment, the temperature of matrix Mg rises swiftly due to the momentary energy deposition, resulting in melting of Mg at the surface layer. As the surface temperature reaches the boiling point, the vaporization goes ahead within a certain range at the surface layer and a lot of droplets sputter out. Because the target material surface parallels to the direction of gravity, the spilled droplets will not return to the surface under the action of gravity but will be charged with negative electricity under the action of electron beam current. As the target is the anode, the droplets with negative electricity will slow down and be drawn back to anode under the action of electric field force. Consequently, the droplets fall on the entire surface at certain speed. After solidification, we can see the pure Mg particles as shown in the Fig. 3.8.

### 3.2.2 Analysis of sectional microstructure

Fig. 3.9 refers to cross sectional SEM image of Mg alloy ZK60-1Y after HCPEB treatment. From the figure, it can be observed that remelted layer with thickness of ~12μm is formed on the surface layer. Compared to the matrix, the remelted layer is more compact and the crystal boundary become less obvious. It can mainly be explained as follows: the second phase existing in crystal boundaries melts and diffuses due to the rapid melting process induced by HCPEB treatment, while the crystal boundaries disappear in the melting liquid. When the electron beam bombardment terminates, the rapid solidification occurs immediately on the surface owing to good thermal conductivity of the metal material. Under the action of high solidification rate (~$10^8$K/s), great temperature gradient is generated on the surface of the material and non-equilibrium freezing occurs in the melting

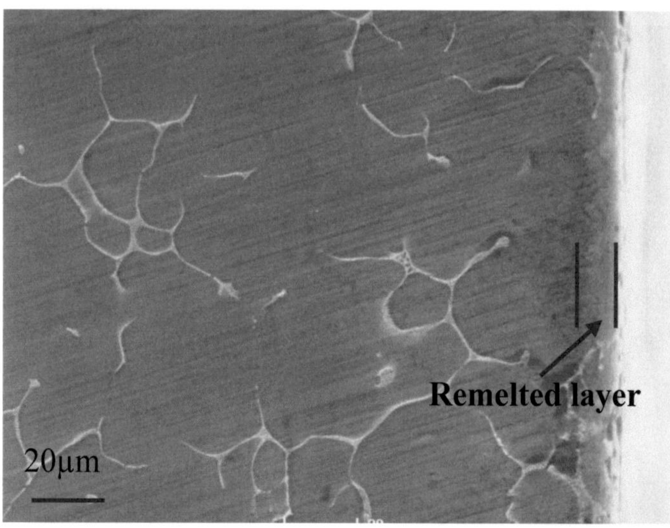


Fig. 3.9. Cross sectional SEM image of Mg alloy ZK60-1Y after HCPEB treatment after HCPEB treatment

layer. For the solidification time is too short, the actual temperature of the liquid at the frontier of solid liquid interface is lower than the liquidus temperature, so there is no way to provide continuous driving force for the growth of crystal grains. The nucleation rate of crystal is greater than the growth rate, so the crystal grains in the surface layer are refined and the inter-grain boundaries of the refined crystal grains become indistinct. In addition, the segregation phenomenon of alloying elements existing in grain boundaries is improved, and Zn and Zr better dissolve into the Mg matrix and form supersaturated solid solution. As a result, the element is uniformly distributed on the modified surface of Mg alloy ZK60-1Y.

### 3.2.3 Friction and wear property of Mg alloy ZK60-1Y before and after HCPEB treatment

Fig. 3.10 indicates the evolution of friction coefficients with the number of friction turns for Mg alloy ZK60-1Y before and after HCPEB treatment. From the figure, it can be seen that the friction coefficients of initial sample maintain around 0.28 at the initial friction stage. However, the friction coefficients of the sample are relatively high at the initial friction stage after 5 pulse treatment, around 0.35. It mainly because the sample surface is not even and the pure Mg particles are distributed on the modified surface, so the surface roughness is increased and the friction coefficients become higher. On the other hand, there is a layer of oxide film on the surface of the initial sample, and the surface roughness is smaller compared to that of the HCPEB-treated sample. Therefore, the friction coefficients of the initial sample are lower at the initial friction stage. With the increase in the number of friction turns, the friction coefficients of the initial sample present rising trend. Upon 1200 cycle friction, the friction coefficients maintain between 0.34 and 0.38. It is because that the oxide layer on the surface is worn away after a period of friction, therefore the matrix gradually touches stainless counterpart discs and shows high friction coefficients.

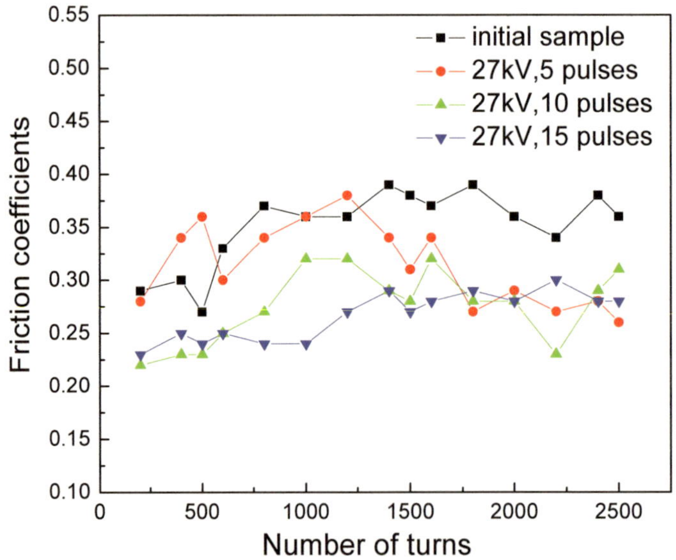


Fig. 3.10. Evolution of friction coefficients with the number of friction turns for Mg alloy ZK60-1Y before and after HCPEB treatment

After 1400 cycle friction, the friction coefficients of the sample treated by HCPEB are lower than that of the initial sample, showing better wear resistance. The reason can be summarized as:1) the grain structure in the remelted layer after HCPEB treatment is refined, namely fine-grain strengthening effect; 2)the enrichment phenomenon of alloying elements is improved, and the chemical composition is distributed evenly and the mechanical property is further enhanced on the ZK60-1Y Mg alloy surface. With the progressing of wear test (after 1800 cycle friction), the friction coefficients of HCPEB treated sample are much lower. It is because the surface at the remelted layer has been worn away basically. Then the remelted layer in this layer has better wear resistance. In the whole friction test, the friction coefficients of the alloy don't fluctuate drastically and maintain between 0.23 and 0.30 after 15 pulse treatment under the accelerating voltage of 27kV. With the increase of pulse number, the times of rapid solidification occurred on the treated sample surface is increased accordingly, thus the alloying elements are distributed evenly and the mechanical property of the alloy surface is more stable. To sum up, compared to the initial sample, the surface of Mg alloy ZK60-1Y has better wear resistance after HCPEB treatment.

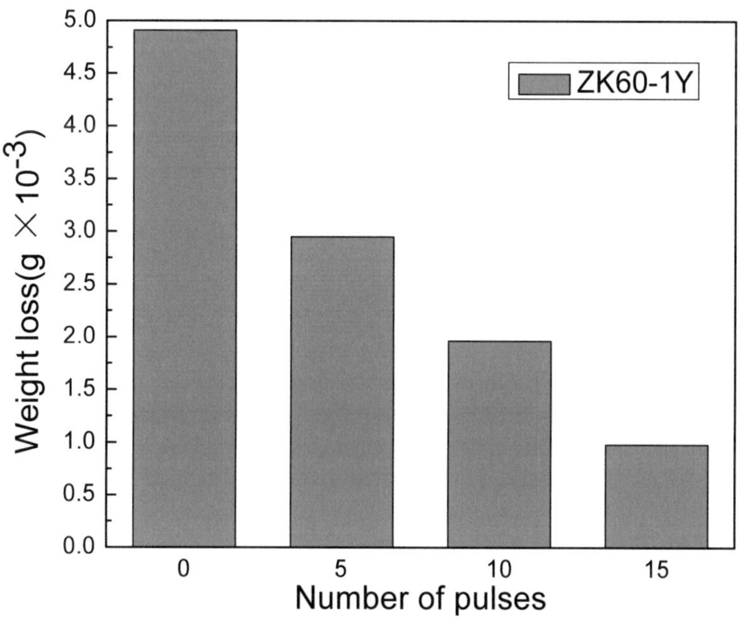


Fig. 3.11. Evolution of weight loss with the number of pulses for Mg alloy ZK60-1Y before and after HCPEB treatment

Fig. 3.11 refers to the evolution of weight loss with the number of pulses for Mg alloy before and after HCPEB treatment. From the figure, it can be seen that the weight loss of the sample is reduced from $4.91 \times 10^{-3}$g of initial sample to $0.98 \times 10^{-3}$g after 15 pulse treatment. It indicates that the wear resistance of the sample by HCPEB treatment is greatly improved under same friction condition. The weight loss of the sample by 5, 10 and 15 pulse treatments is reduced by 39.9%, 60.1%, 80% respectively compared to that of the initial sample. That means the wear resistance of the sample is the best after 15 pulse treatment under the accelerating voltage of 27kV and the relative wear resistance is improved by a factor of 5. It can be attributed to the changes in the microstructure of ZK60-1Y Mg alloy surface after HCPEB treatment, namely composition homogenization (segregation reduction), grain refinement and the formation of supersaturated solid solution.

## 4. Conclusions

For the first time, the HCPEB technology is utilized to treat the surface of Mg alloys AZ31 and ZK60-1Y in the paper and perform the research into the changes in microstructure of Mg alloys before and after HCPEB treatment and the effect on the friction and wear property and corrosion resistance. The main conclusions are listed as below:

**1) Mg alloy AZ31**

After HCPEB irradiation of Mg alloy AZ31, wavy and "crater" morphologies and twins emerge on the modified surface. According to XRD analysis, it indicates that the diffraction peaks of Mg obviously move towards high-angle direction due to stresses existing in the surface layer induced by rapid solidification during HCPEB process.

After HCPEB treatment, the friction coefficients of Mg alloy AZ31 are reduced and the wear resistance is greatly improved. The wear resistance of the 15-pulsed sample is enhanced by about a factor of 6. Besides, the wear form is the typical abrasive wear. Meanwhile, the result of polarization curve test in 5%NaCl indicates that the corrosion resistance of the sample after HCPEB treatment is enhanced because the content of Al on the alloy surface is increased, and a layer of compact oxide film is formed on the surface. As a result, the polarization resistance is increased and the corrosion current is reduced.

**2) Mg alloy ZK60-1Y**

During the HCPEB treatment, the rapid melting and solidification process is completed on the surface of Mg alloy ZK60-1Y, and the grain boundaries on the surface become less obvious. The chemical composition tends to uniform distribution and some pure Mg particles are distributed on the alloy surface. According to the analysis of cross sectional structure, it indicates that a remelted layer with the thickness of ~12μm is formed on the surface of Mg alloy ZK60-1Y, which possesses much compact and refined structure compared to the matrix.

According to friction and wear test, the wear resistance of Mg alloy ZK60-1Y surface is obviously enhanced after HCPEB treatment and the friction coefficients and weight loss are reduced compared to that of the original sample. The wear resistance of Mg alloy ZK60-1Y by 15 pulse treatment is the best and the relative wear resistance is increased by a factor of 5.

## 5. Acknowledgements

This study has been supported by the Fundamental Research Funds for the Central Universities (N090602009 & N100402010), the Key Projects in the National Science & Technology Pillar Program during the Eleventh Five-Year Plan Period (2009BAE80B01&2009BAE85B03), and the Education Department Project of Liaoning Province of China (2008T239 & LT2010036)

## 6. References

[1] Liu X.B., Shan D.Y., Song Y.W. & Han E.H. (2010). Effects of heat treatment on corrosion behaviors of Mg-3Zn magnesium alloy, *Transactions of Nonferrous Metals Society of China* Vol.20 (No.7): 1345-1350.

[2] Feliu S. J., Pardo A., Merino M. C., Coy A. E., Viejo F. & Arrabal R. (2009). Correlation between the surface chemistry and the atmospheric corrosion of AZ31, AZ80 and AZ91D magnesium alloys, *Applied Surface Science* Vol.255 (No.7): 4102-4108.

[3] Candan S., Unal M., Koc E., Turen Y. & Candan E. (2011). Effects of titanium addition on mechanical and corrosion behaviours of AZ91 magnesium alloy, *Journal of Alloys and Compounds* Vol.509 (No.5): 1958-1963.

[4] Potzies C. & Kainer K.U. (2004). Fatigue of magnesium alloys, *Advanced Engineering Materials* Vol.6 (No.5): 281-289.

[5] Bin Sajuri Z., Umehara T., Miyashita Y. & Mutoh Y. (2004). Fatigue-Life Prediction of Magnesium Alloys for Structural Applications, *Advanced Engineering Materials* Vol.5 (No.12): 910-916.

[6] Song G.L. & Atrens A. (1999). Corrosion mechanisms of magnesium alloys, *Advanced Engineering Materials* Vol.1 (No.1): 11-33.

[7] Wu K., Wang Y.Q. & Zheng M.Y. (2007). Effects of microarc oxidation surface treatment on the mechanical properties of Mg alloy and Mg matrix composites, *Materials Science and Engineering A* Vol.447 (No.1-2): 227-232.

[8] Chen C.J., Wang M.C., Wang D.S., Jin R. & Liu Y.M. (2007). Microstructure and corrosion behavior of Mg-Nd coatings on AZ31 magnesium alloy produced by high-energy micro-arc alloying process, *Journal of Alloys and Compounds* Vol.438 (No.1-2): 321-326.

[9] Jiang B.L., Zhang S.F., Wu G.J. & Lei T.Q. (2002). Microflaw and phases constitution of ceramic coating formed by micro-arc oxidation on magnesium alloys and their influence on corrosion-resistance, *The Chinese Jouranl Nonferrous metals* Vol.12 (No.3): 454-457.

[10] Apelfeld A. V., Bespalova O. V., Borisov A. M., Dunkin O. N., Goryaga N. G., Kulikauskas V. S., Romanovsky E. A., Semenov S. V. & Souminov I. V. (2000). Application of the particle backscattering methods for the study of new oxide protective coatings at the surface of Al and Mg alloys, *Nuclear Instruments and Methods in Physics Research B* Vol.161-163: 553-557.

[11] Wang Y.M., Wang F.H., Xu M.J., Zhao B., Guo L.X. & Ouyang J.H.(2009). Microstructure and corrosion behavior of coated AZ91 alloy by microarc oxidation for biomedical application, *Applied Surface Science* Vol.255 (No.22): 9124-9131.

[12] Abbas G., Liu Z.& Skeldon P. (2005). Corrosion behaviour of laser-melted magnesium alloys, *Applied Surface Science* Vol.247 (No.1-4): 347-353 .

[13] Dutta Majumdar J., Galun R., Mordike B. L. & Manna I. (2003). Effect of laser surface melting on corrosion and wear resistance of a commercial magnesium alloy, *Materials Science and EngineeringA* Vol.361 (No.1-2): 119-129 .

[14] Samant Anoop N., Du Baoshuai, Paital Sameer R., Kumar S. & Dahotre Narendra B.(2009). Pulsed laser surface treatment of magnesium alloy: Correlation between thermal model and experimental observations, *Journal of Materials Processing Technology* Vol.209 (No.11): 5060-5067.

[15] Mondal A.K., Kumar S., Blawert C. & Dahotre Narendra B. (2008). Effect of laser surface treatment on corrosion and wear resistance of ACM720 Mg alloy, *Surface and Coatings Technology* Vol.202 (No.14): 3187-3198.

[16] Dutta Majumdar J. & Manna I.(2010). Mechanical properties of a laser-surface-alloyed magnesium-based alloy (AZ91) with nickel, *Scripta Materialia* Vol.62 (No.8): 579-581.

[17] Bohne Y., Seeger D.M., Blawert C., Dietzel W., Mändl S. & Rauschenbach B. (2006). Influence of ion energy on properties of Mg alloy thin films formed by ion beam sputter deposition, *Surface and Coatings Technology* Vol.200 (No.22-23): 6527-6532.

[18] Li P., Lei M.K., Zhu X.P., Han X.G., Liu C. & Xin J.P. (2010). Wear mechanism of AZ31 magnesium alloy irradiated by high-intensity pulsed ion beam, *Surface and Coatings Technology* Vol.204 (No.14): 2152-2158.

[19] Yang J.X., Cui F.Z., Lee In-Seop, Jiao Y.P., Yin Q.S. & Zhang Y. (2008). Ion-beam assisted deposited C-N coating on magnesium alloys, *Surface and Coatings Technology* Vol.202 (No.22-23): 5737-5741.

[20] Zou J.X., Grosdidier T., Zhang K.M. & Dong C. (2009). Cross-sectional analysis of the graded microstructure in an AISI D2-steel treated with low energy high-current pulsed electron beam, *Applied Surface Science* Vol.255 (No.9): 4758-4764.

[21] Hao S.Z, Zhang X.D., Mei X.X., Grosdidier T. & Dong C.(2008). Surface treatment of DZ4 directionally solidified nickel-based superalloy by high current pulsed electron beam, *Materials Letters* Vol.62 (No.3): 414-417.

[22] Dong C., Wu A., Hao S., Zou J., Liu Z., Zhong P., Zhang A., Xu T., Chen J., Xu J., Liu Q. & Zhou Z.(2003). Surface treatment by high current pulsed electron beam, *Surface and Coatings Technology* Vol.163-164: 620-624.

[23] Surzhikov A.P., Frangulyan T.S., Ghyngazov S.A. & Koval N.N. (2009). Structural-phase transformations in near-surface layers of alumina-zirconium ceramics induced by low-energy high-current electron beams, *Nuclear Instruments and Methods in Physics Research B* Vol.267 (No.7): 1072-1076.

[24] An J., Shen X.X., Lu Y. & Liu Y.B.(2006). Microstructure and tribological properties of Al-Pb alloy modified by high current pulsed electron beam, *Wear* Vol.261 (No.2): 208-215.

[25] Hao Y., Gao B., Tu G.F., Li S.W., Hao S.Z. & Dong C.(2011). Surface modification of Al-20Si alloy by high current pulsed electron, *Applied Surface Science* Vol.257 (No.9): 3913-3919.

[26] Zhang K. M., Zou J. X., Grosdidier T., Gey N., Weber S., Yang D. Z. & Dong C.(2007). Mechanisms of structural evolutions associated with the high current pulsed electron beam treatment of a NiTi shape memory alloy, *Journal of Vacuum Science and Technology A* Vol.25 (No.1): 28-36.

[27] Proskurovsky D.I., Rotshtein V. P., Ozur G. E., Markov A. B., Nazarov D. S., Shulov V. A., Ivanov Yu. F. & Buchheit R. G. (1998). Pulsed electron-beam technology for surface modification of metallic materials, *Journal of Vacuum Science and Technology A* Vol.16 (No.4): 2480-2488.

[28] Guan Q.F., Zhang Q.Y., Dong C. & Zou D.T. (2005). Deformation twining in single-crystal aluminum induced by high-current pulsed electron beam, *Journal of Mterials Science* Vol.40 (No.18): 5049-5052.

[29] Gao B., Hao S.Z., Zou J. X., Wu W. Y, Tu G.F. & Dong C. (2007). Effect of high current pulsed electron beam treatment on surface microstructure and wear and corrosion resistance of an AZ91HP magnesium alloy, *Surface and Coatings Technology* Vol.201 (No. 14): 6297-6303.

[30] Mizutani Y., Kim S. J., Ichino R. & Okido M. (2003). Anodizing of Mg alloys in alkaline solutions, *Surface and Coatings Technology* Vol.169-170: 143-146.

[31] Qin Y., Dong C., Wang X.G., Hao S.Z., Wu A.M., Zou J.X. & Liu Y. (2003). Temperature profile and crater formation induced in high current pulsed electron beam processing, *Vacuum Science and Technology A* Vol.21 (No.6): 1934-1938.

# 3

# Estimation of Carbon Coatings Manufactured on Magnesium Alloys

Marcin Golabczak
*Technical University of Lodz,*
*Department of Production Engineering*
*Poland*

## 1. Introduction

Magnesium (Mg) is one of the most abundant structural metals on the earth. Magnesium resources are estimated on approximately 1,93% of mass of the earth's crust and 0,13% of mass of the oceans. Magnesium is present in salt water in form of chlorides (in amount of approximately 1,2 kg/m$^3$), however in earth crust in form of dolomites composed mainly from carbonates. Magnesium belongs to ultra light metals (1,75 g/cm$^3$), has silver glossy colour, is soft and ductile, easily reacts chemically with other substances (e.g.: oxygen, nitrogen, carbon dioxide or water). Unfortunately, magnesium has a lot of undesirable properties such as poor corrosion and wear resistance, what limits its use in many usages especially for outdoor applications. For this reason pure magnesium is rarely used in technique, however with other metals (e.g.: aluminum, zinc, manganese, cerium, zirconium and rare earth metals) forms alloys, which are very attractive constructional material. Because of this magnesium alloys found a plethora of applications in various branches of industry where reduction in weight is of importance (Gray & Luan, 2002). These alloys are used in aerospace, automobile (Kawalla et al., 2008) and electronic industries, for manufacturing of sporting goods, high-speed boats, submarines, household equipment, etc. (Fig. 1). The main advantages of magnesium alloys are: the high strength, weight ratio, high thermal conductivity, small heat extensibility, good welding characteristics and high functional integrity, which allow to produce near-net-shape elements as well as good machinability (Hawkins, 1993). However, magnesium alloys have also certain disadvantages. The most troublesome of them is the high susceptibility to corrosion (especially galvanic corrosion), which contributes to dwindling of their size and reduces mechanical durability. As to protect magnesium alloys from corrosion, at present various methods for the fabrication of protective films have been used (Ishizaki et al., 2009). Other disadvantages of magnesium alloys comprise their weak wear resistance, a drop in durability at high temperature and interference of electromagnetic field. The aforementioned faults considerably reduce the area of application of this material. Presented studies aimed at elimination of the listed drawbacks by means of covering of magnesium alloy with special carbon coatings. Plasma Activated Chemical Vapor Deposition (PACVD) method has been used for this purpose. Optimum conditions of this process have been determined and the material properties of the carbon coatings characterized.

a)

b)

c)

d)

Fig. 1. Examples of use magnesium alloys: a) mobile phone housing, b) wheelcase, cover and flange of differential gear, c) magnesium alloy wheels, or "mag wheels" used on racing cars, d) body front of camera

## 2. Methods of manufacturing of protective coatings on magnesium alloys

In literature there are many different methods and techniques of manufacturing of protective and decorative coatings on magnesium alloys (Gray & Luan, 2002). Taking into consideration physical processes used in these techniques we can classify seven main methods of manufacturing of protective and decorative coatings on magnesium alloys, which have been presented in figure 2. These include electrochemical plating, conversion coatings, hydride coatings, anodizing, vapour-phase processes, laser cladding and polymer coatings (Golabczak, 2005). All these methods are characterized by different complexity of used technological processes, costs of realization these processes, degree of environmental nuisance and surroundings, as well as functional properties of manufactured coatings and range of their applications. As yet it has not been developed effective method assuring complete corrosion resistance of magnesium alloys and decorative virtues of manufactured coatings. Because of the increasing interest in magnesium alloys in different fields of industry, it is justified to carry out research on elaboration of a "new" methods fulfilling all these requirements in superlative degree.

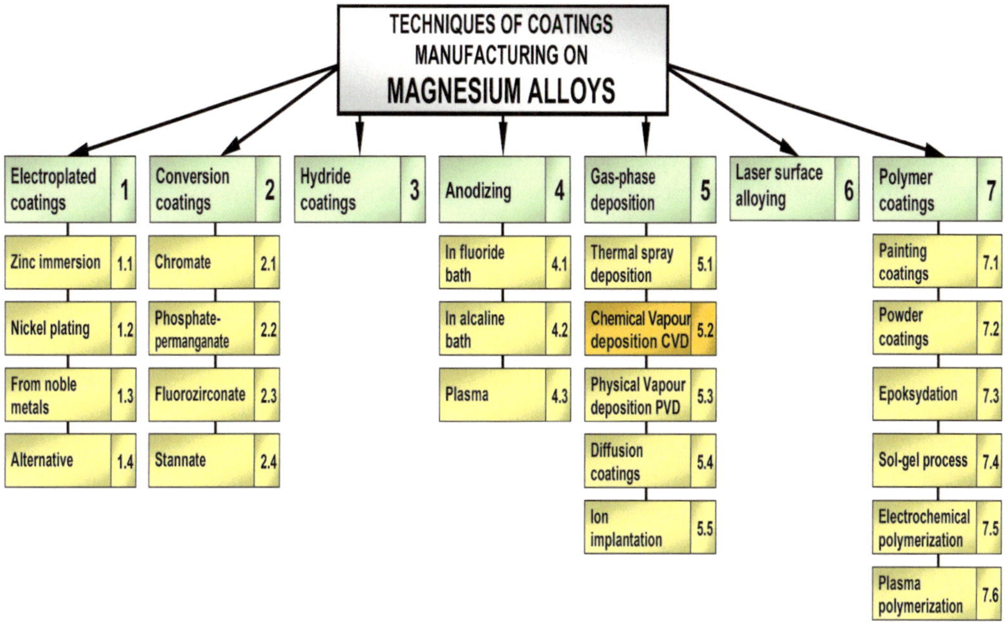


Fig. 2. Techniques of coatings manufacturing on magnesium alloys

## 3. Characteristic of carbon coatings

Carbon coatings have been characterized by very attractive functional properties, especially by decorative and protective, which predestine them for application in many fields. At present, there are many methods and techniques which are used for their manufacturing, among them dominate techniques exploitative plasma, ion beams and methods of unconventional synthesis (Robertson, 2002). Diversity of these methods and wide range of applied parameters have essential influence on quality of manufactured carbon coatings. Taking into account the structure of manufactured carbon coatings we can identify four basic groups:

- **diamond** - inclusive diamond films - DF and polycrystalline diamond coatings - PCD, which are composed of atoms of configuration $\sigma sp^3$, nanocrystalline diamond coatings - NCD, tetrahedral carbon ta-C and amorphous diamond a-D coatings;
- **graphite** - amorphous carbon coatings of graphite structure e.g. *pyrolytic graphite coatings* which are obtained in vacuum pyrolysis process;
- **carbyne** - inclusive *α-carbyne contains*, which contains acetylic bonds (-C≡C-) and are also called as *polyacetynele carbyne contains*, *β-carbyne coatings*, which contains cumulative double bond (=C=C=) also called as *polycumulene carbyne coatings;*
- **diamond like carbon** - inclusive *diamond like carbon coatings* – DLC – which are mixture of amorphous or nanocrystalline of carbon containing fraction of $\sigma sp^3$ bonds - typical for diamond structure, fraction of $\sigma sp^2$ bonds – typical for graphite and $\sigma sp^1$.

Above mentioned structures of carbon coatings have found application for deposition of many constructional materials used for example in: medicine for manufacturing of implants (Niedzielski et al., 1997), tool industry for increasing of durability and wear resistance of cutting edges (Olszyna & Smolik, 2004), jewellery industry for manufacturing of decorative coatings (Clapa et al., 2001), for coating of polymers used in aerospace industry (Hawkins, 1993), etc. This wide field of application of carbon coatings justifies usefulness their exploitation for covering of magnesium alloys. Among many analyzed methods of manufacturing of carbon coatings especially attractive seems to be PACVD (Plasma Activated Chemical Vapour Deposition) method, elaborated in Technical University of Lodz - Poland. This method is particularly useful for manufacturing carbon coatings mentioned above with predominated part of diamond in these coatings (Niedzielski et al., 1997).

## 4. The stand for manufacturing of carbon coatings by PACVD method

For investigations AZ31 magnesium alloy samples have been used. AZ31 (ASTM designation) is very commertial alloy used in die casting and plastic forming. The chemical composition (in wt%) of AZ31 is: 2.83 %Al, 0.8% Zn, 0.37% Mn and 0.002% Cu (Kuc et al., 2008). Carbon coatings have been deposited on this alloy by PACVD method, which has relied on decomposition of methane in electric field with high frequency of 13.56 MHz, obtained at the pressure of approximately 12 Pa in a working chamber (Golabczak, 2005). Processes of PACVD have been realized in the stand presented in figure 3. It has consisted of the chamber of water cooled plasma reactor, the high frequency electrode fixed to the plate of the base and connected through the condenser (the latter provided the negative potential of self-polarization), generator of high frequency (facilitated production of plasma with high density and maintained the frequency at the constant level), vacuum system and systems of measurement and control. Carbon coatings have been deposited on AZ31 magnesium alloy in two steps comprising the process of ionic digestion of their surface followed by the process of synthesis of these coatings. Parameters of these steps are shown in table 1.

Fig. 3. The view of the system used for deposition of carbon coatings by PACVD method

| Parameter | Ionic digestion of the surface | Process of coating deposition |
|---|---|---|
| Feed gas | $CH_4$ | $CH_4$ |
| Pressure in a working chamber | 8 ÷ 10 Pa | 12 Pa |
| Time of process - t | 4 min | 5 ÷ 9 min |
| Gas flow rate - V | 5 $cm^3/min$ | 20÷60 $cm^3/min$ |

Table 1. Optimum parameters of PACVD process

## 5. Preparation of samples made of magnesium alloys for investigations

Polishing process of samples made of AZ31 magnesium alloy has been carried out using Phoenix Beta 2 (Buehler-Germany) dual platen grinder-polisher machine equipped with Vector power head (Fig. 4) and specimens holder for single force for 3-6 specimens up to max ø 25 mm, according to holder selected. Thus 3 to 6 specimens can be prepared under reproducible conditions. The Buehler grinder-polisher machine has had stepless rotation speed (from 30 to 600 rpm) and the power head settings of control time, pressure (up to 200 N), speed and direction and automatic start and stop system. Vector power head upgrades the Beta 2 grinder-polisher machine to from manual operation to semi-automatic operation, increasing productivity and specimen consistency. This stand is on equipment of Department of Production Engineering of Technical University of Lodz - Poland laboratory.

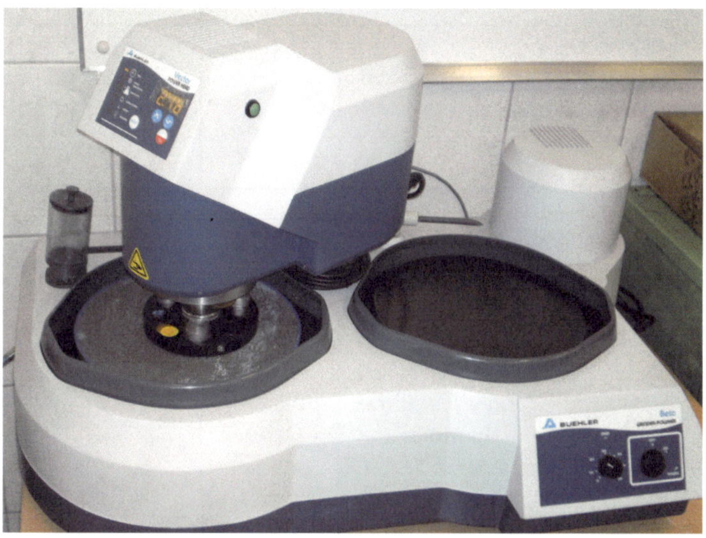


Fig. 4. The overall view of the Beta 2 dual platen grinder-polisher machine equipped with Vector power head and specimens holder

The samples made of AZ31 hp magnesium alloy of diameter ø 20 mm from TECHNO-COAT Oberflaechentechnik GmbH, Zittau-Germany have been used for investigations. The main technological requirement of technological process was to prepare samples of a low surface roughness and removal impurities from their surface layer. For this purpose the technological process of abrasive machining including following grinding and polishing operations, using Buehler equipment and accessories has been elaborated:

- two stage grinding of samples on grinder equipped with self adhesive BuehlerMet silicon carbide abrasive paper; in sequential stages of grinding the granularity of SiC material has been diversified using accordingly: in first stage silicon carbide grits size 26 μm and in second stage silicon carbide grits size 26 μm; the grinding process has been carried out using wax;
- lapping of samples using medium hard woven silk cloth VerduTex and Buehler MetaDi diamond suspension of diamond grains size 3 μm; MetaDi is a oil-base product, particularly suitable for soft and water-sensitive materials, absolutely water free with tight distribution of synthetic, monocrystalline diamonds which have a great number of cutting faces. This offers a particularly high material removal rate and scratch-free surfaces of samples;
- final polishing of samples using self-adhesive soft synthetic pad ChemoMet and aluminium oxide ($Al_2O_3$) final polishing suspension MasterPrep (grains size ø 0,05 μm);
- washing of samples in ethyl alcohol of high purity (99,9 %) Chem Land using ultrasonic washer Polsonic – Sonic 1.

The technological conditions of realized operations of grinding and polishing of AZ31 magnesium alloy samples have been shown in table 2.

| Process stages | Abrasive surface | Type of abrasive material | Lubricant type | Process time [min] | Feed force [N/cm$^2$] | Rotation speed of platen V [m/s] |
|---|---|---|---|---|---|---|
| Grinding of sample surface | BuehlerMet silicon carbide abrasive paper | Silicon carbide SiC P 600 (grits size ø 26 μm) | Wax | 5 | 10 | 6 |
| | BuehlerMet silicon carbide abrasive paper | Silicon carbide SiC P 1200 (grits size ø 15 μm) | Wax | 1 | 5 | 3 |
| Lapping | Medium hard woven silkcloth VerduTex | Monocrystalline diamond suspension MetaDi -oil based (grains size ø 3 μm) | Oil-based polishing extander Buehler AutoMet Lapping Oil | 5 | 2.5 | 2 |
| Polishing | Soft synthetic pad ChemoMet | Aluminium oxide ($Al_2O_3$) final polishing suspension MasterPrep (grains size ø 0,05 μm) | - | 3 | 2,5 | 1 |

Table 2. Conditions of technological process of magnesium alloy samples preparation

Elaborated technological process has ensured suitable preparation of samples, in range of required roughness parameters and their proper purity (including removal of machining products). The surface of polished samples has had silver, glossy colour and no visible tool marks.

## 6. Experimental investigations

Experimental investigations have included determination of technological parameters of the process of manufacturing of carbon coatings on AZ31 magnesium alloy using PACVD method and test of operational properties of coatings manufactured in this process. For arrangement of technological parameters of PACVD process the investigation object model has been accepted, on which affects variable input quantities (controllable), constant quantities and factors of jamming. Modelling setting-up of investigation conditions, inclusive of: variable and constant input quantities, factors of jamming of PACVD process and output quantities have been presented in figure 5.

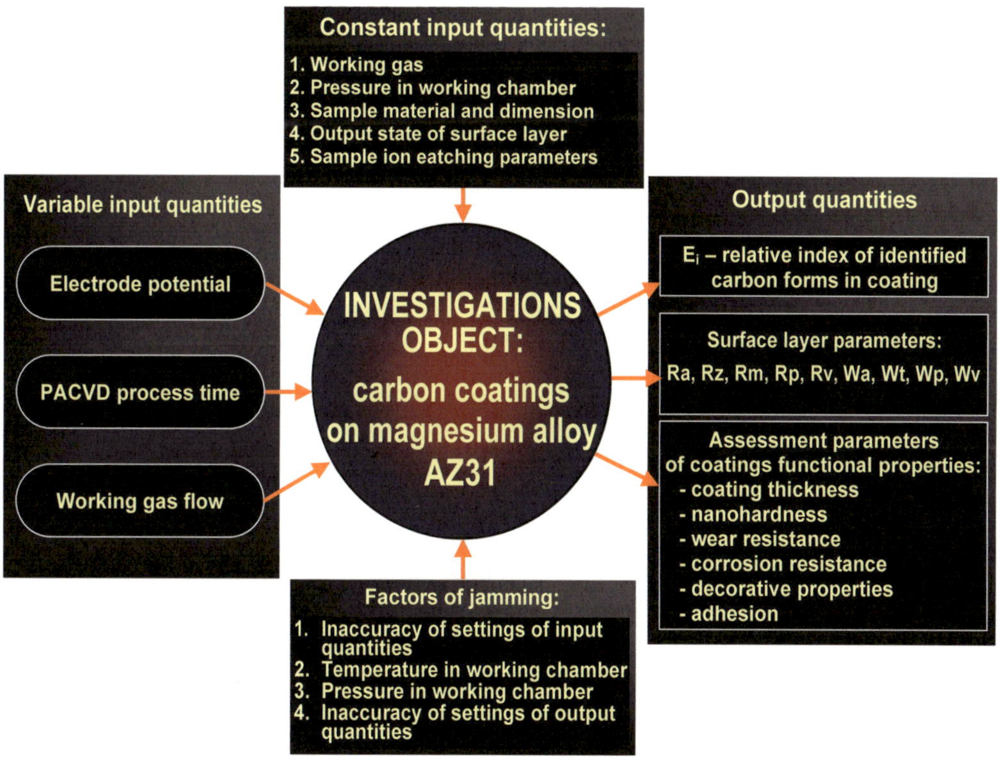


Fig. 5. Modelling setting-up of investigation conditions

Input quantities and areas and their range of variables have been determined basing on technological possibilities of test stand for manufacturing of carbon coatings using PACVD method and preliminary investigation results (Golabczak, 2005). Results of preliminary tests have shown, that polarization potential values of high frequency electrode should not exceed voltage of 1000 V and time of PACVD process 12 minutes. Overdraft of these values has caused excessive heating of samples made of magnesium alloy and their burning. As output quantities the set of parameters useful both for identification of manufactured carbon coatings on magnesium alloy samples in PACVD process and assessment of their functional properties (Fig. 5). Determined investigation conditions, range of variables and test step have been shown in table 3.

| Range of variables and input quantities of PACVD process | | | |
|---|---|---|---|
| Input quantities / Range of variables and test step | Polarization potenctial U [V] | Time of PACVD process t [min] | Working gas flow V [cm³/min] |
| Ground level | 800 | 7 | 40 |
| Test step | 100 | 2 | 20 |
| Upper level | 900 | 9 | 60 |
| Lower level | 700 | 5 | 20 |
| **Constant input quantities** | | | |
| No. | Name of quantity | Value/determination | |
| 1 | Working gas -methane | $CH_4$ | |
| 2 | Pressure in working chamber | 12 Pa | |
| 3 | Samples dimension | Ø 20x2 mm | |
| 4 | Preparation of sample surface layer | According to procedure depicted in point No. 5 | |

Table 3. Investigation conditions and range of variables of input quantities of manufacturing process of carbon coatings using PACVD method

Taking into consideration limitation of costly and labour-intensive experiment designs, the planned fractional experiment $2^{n-1}$ type has been accepted, in which number of experiments has been equal to 4 (Golabczak, 2005). Design matrix for this type of experiment has been presented in table 4.

| **No. of sample** | **Variable input quantities** | | |
|---|---|---|---|
| | **$X_1$ Polarization potenctial of electrode U [V]** | **$X_2$ Time of PACVD process t [min]** | **$X_3$ Working gas flow V [cm³/min]** |
| 1 | + | – | – |
| 2 | – | + | – |
| 3 | + | – | + |
| 4 | – | + | + |
| + upper level of input variable (according to data in table 3)<br>– ground level of input variable (according to data in table 3) | | | |

Table 4. Plan of fractional experiment $2^{n-1}$ type of manufacturing of carbon coatings

## 7. Experimental results

Experiments have included optimization of technological parameters of PACVD process of deposition of carbon films and characterization of their material properties. To determine the optimum conditions of PACVD process, series of tests have been realized according to the fractional experiment $2^{n-1}$ type (Golabczak, 2005). The studies of material properties of carbon coatings deposited on AZ31 magnesium alloy have comprised identification of these coatings, measurement of their nanohardness and thickness and determination of their geometrical microstructure of surface. Also tribologic measurements of hard carbon coatings, determination of their corrosion resistance and adhesion have been realized.

### 7.1 Identification of carbon coatings deposited on AZ 31 magnesium alloy

Identification of hard carbon coatings deposited on the surface of AZ31 magnesium alloy has been determined on the basis of their Raman spectra (Golabczak, 2005). For this purpose, the mathematical modeling of fitting of Gaussian peaks of the identified carbon phases to the Raman spectra has been performed (Golabczak, 2005). Exemplary results of Raman spectrum evaluation with using Gaussian profile have been shown in figure 6.

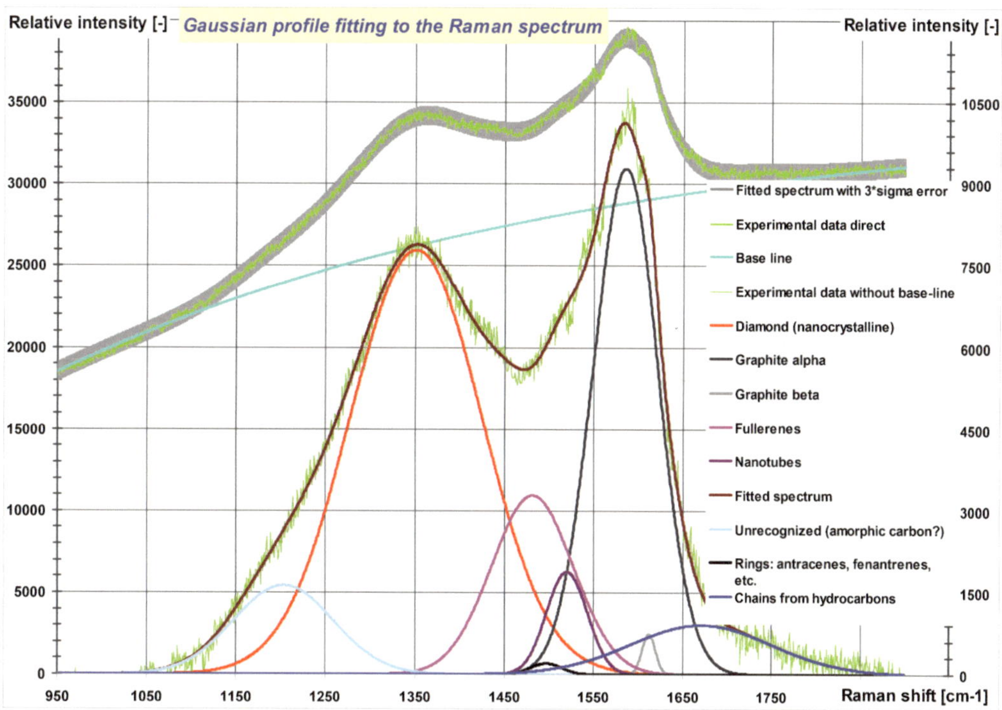


Fig. 6. Results of Raman spectrum evaluation with using Gaussian profile

The contents of the identified carbon phases in the deposited carbon coatings have been determined on the basis of the relative index - $E_i$, described by the following equation (1):

$$E_i = \frac{A_{p_i}}{A_s} = \frac{\int p_i(x)dx}{\sum \int p_i(x)dx} \tag{1}$$

where: $A_{pi}$ - surface area between the baseline and the curve of fitting to the plot for the carbon phases identified in the coating, calculated by the method of numerical integration; $A_s$ - the summary surface area between the baseline and the curve of fitting to Raman spectrum, calculated by the method of numerical integration (Golabczak, 2007). The quality of identified carbon forms, found in coatings deposited through the successive experimental trials has been determined on the basis of the accepted mathematical model used for fitting the Gaussian peaks spectra to Raman spectra. For this purpose, the relative index $E_i$ has been calculated (according to equation 1). The relative index $E_i$ determines the content of individual carbon forms in the deposited coatings. The graphical interpretation of this method has been shown in figure 7, which presents the relative content of diamond in the deposited carbon coating, attained under optimum PACVD process conditions.

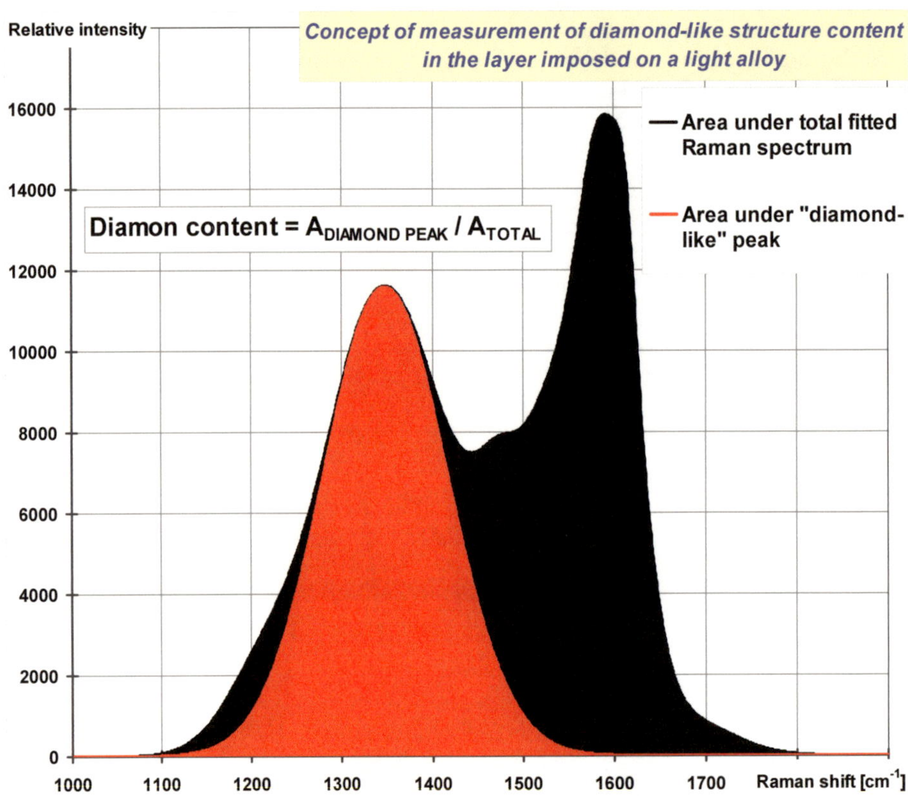


Fig. 7. The concept of measurement of the "diamond-like" structure content in the carbon coating on a light alloy

Results of computing of the relative index $E_i$, obtained for each specimen prepared within the scope of the planned experiment, have been shown in figure 8.

Analysis of results obtained for individual specimens produced within the scope of the planned experiment (specimens 1÷4), revealed that the diamond phase ($E_i$=0.40÷0.48) and alfa-graphite phase ($E_i$=0.20÷0.35) have dominated in the deposited coatings. Also other carbon phases have been identified, such as beta-graphite- ($E_i$=0.003÷0.013), fullerens ($E_i$=0.11÷0.22), nanotubes ($E_i$ of approximately 0.034) and other, including some unidentified forms ($E_i$=0.006÷0.068), rings ($E_i$ of approximately 0.008), and chains ($E_i$=0.018÷0.12), but their contents have been minor. Optimization of deposition conditions (specimen 5) has shown that the rise in contents of diamond phase (to $E_i$=0.54) in the coatings has been achievable. The latter content of diamond phase in the coating has been reached under the following PACVD process conditions: U=900V, t=8min, and V=60cm$^3$/min.

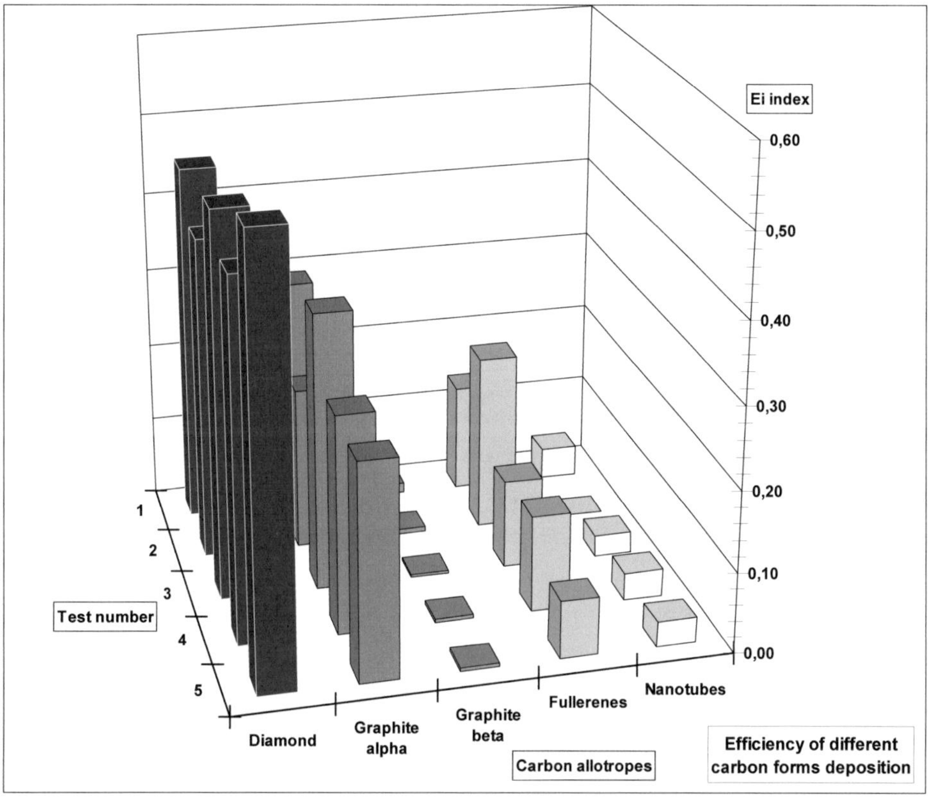


Fig. 8. Comparison of values of the relative index $E_i$ for carbon phases identified in carbon coatings deposited on magnesium alloy specimens obtained within the scope of the planned experiments of PACVD process

## 7.2 Measurements of nanohardness of carbon coatings deposited on AZ31 magnesium alloy

Nanohardness of carbon coatings has been measured using Nano Test 600 meter (Micro Materials Ltd., Great Britain) equipped with a diamond pyramidal penetrator (Golabczak, 2005). The measurements of nanohardness have been conducted at the penetrating force (F) of 0.1–0.6 mN (extorted by the penetrator) and the rate of F increase (dF/dt) of 0.02 mN/s. Values of nanohardness of hard carbon coatings, measured by using the pyramidal penetrator, have been calculated as follows (2):

$$H_n = \frac{F}{24{,}5 h_p^{\,2}} \tag{2}$$

where: $H_n$ - nanohardness of the outer layer [GPa], F - the penetrating force [N], $h_p$ - indentation made by the penetrator [m]. Representative results of nanohardness measurements of the examined hard carbon coatings deposited on AZ31 magnesium alloy have been shown in figure 9.

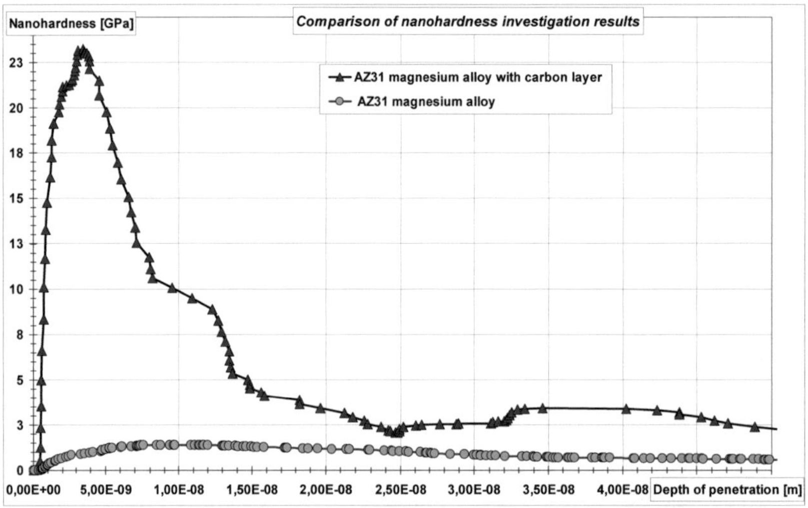


Fig. 9. Comparison of the nanohardness of AZ31 magnesium alloy protected by the carbon coating deposited under optimum conditions of PACVD process and the specimen without this coating

The latter presents the difference between the nanohardness of AZ31 magnesium alloy protected by the carbon coating deposited under optimum conditions of PACVD process and the alloy without this coating. These results provide evidence that the nanohardness of AZ31 magnesium alloy protected by the carbon coating deposited by the PACVD method has been considerably higher (24 GPa) than that of the alloy without the coating (0.8 GPa).

### 7.3 Determination of thickness of carbon coating

The thickness of hard carbon coatings has been determined by the method of direct profilography using the highly precise Taylor Hobson profilographometer (Golabczak, 2005, 2010). To achieve the accurate results of measurements, the carbon coatings have been deposited only on selected fragments of the examined samples of magnesium alloy. Therefore, some parts of their surface have been protected by quartz plates during synthesis of the coatings (Fig. 10). Thus the examined surfaces of magnesium alloy has contained the fragments coated by the carbon coating and free from the latter. Results of these measurements are collected in figure 11. The mean value (from 5 distinct measurements) of the thickness of a carbon coating has been approximately equal to 220 nm.

### 7.4 Determination of geometrical microstructure of surface and morphology of surface layer of carbon coatings

Presented investigation results concern comparison of geometrical microstructure parameters of surface and morphology of carbon coating, manufactured on AZ31 magnesium alloy samples using PACVD method, with analogous parameters of samples

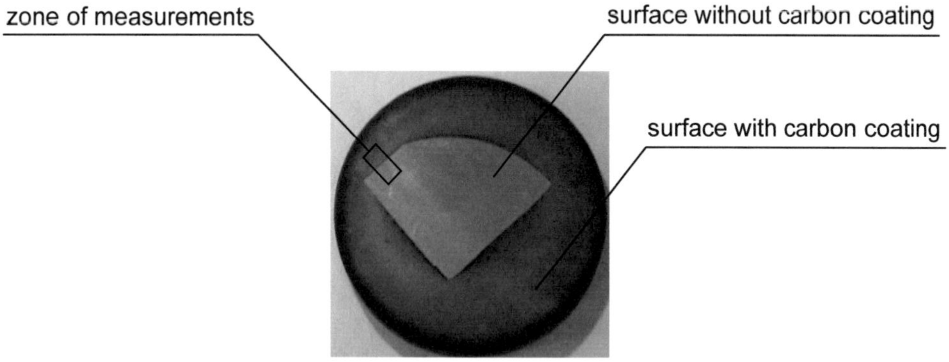


Fig. 10. The image of the surface of a specimen of magnesium alloy AZ31 prepared for the measurements of carbon coating thickness and the zone of measurements carried out by profilography method

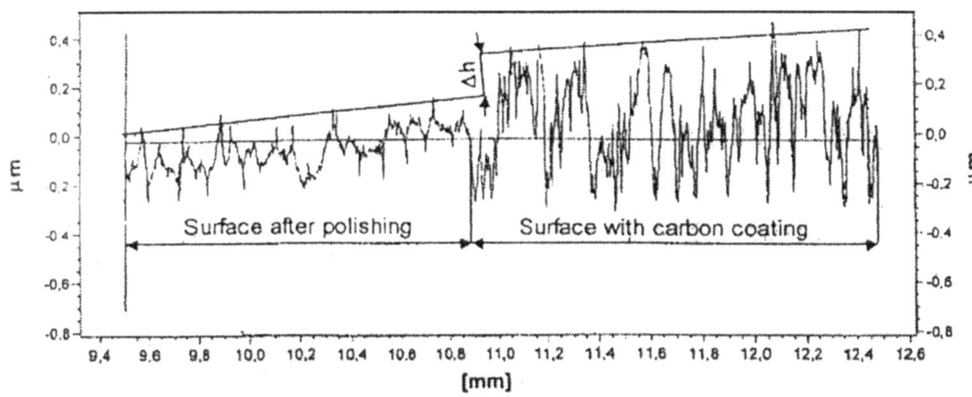


Fig. 11. Results of measurements of the thickness of carbon coating deposited on magnesium alloy AZ31 carried out by the method of direct profilography; the thickness of carbon coating (Δh) of 220 nm

without carbon coating, prepared for deposition process (after polishing process). Investigation range has included estimation of carbon coatings manufactured in optimum conditions of PACVD process (sample No. 5). Geometrical microstructure of carbon coating surface has been estimated basing on roughness and waviness parameters of surface in 2D and 3D configuration, however morphology of carbon coating basing on measurements of atomic force microscope - AFM and images of scanning electron microscope - SEM. The profile measurements have been carried out in Department of Production Engineering of Technical University of Lodz laboratory, using profilometer type PGM-1C IOS. Samples for profile measurements have been prepared in the same way like in case of thickness measurement of carbon coating (point No. 7.3). It has ensured objectivity of measurement results. Exemplary profile measurements results of samples in 2D configuration have been presented in figure 12 and 14, however in 3D configuration in figure 13 and 15. The values of roughness and waviness parameters of samples surfaces have been placed in suitable profilograms and have referred to their average value from five tests of profile measurement.

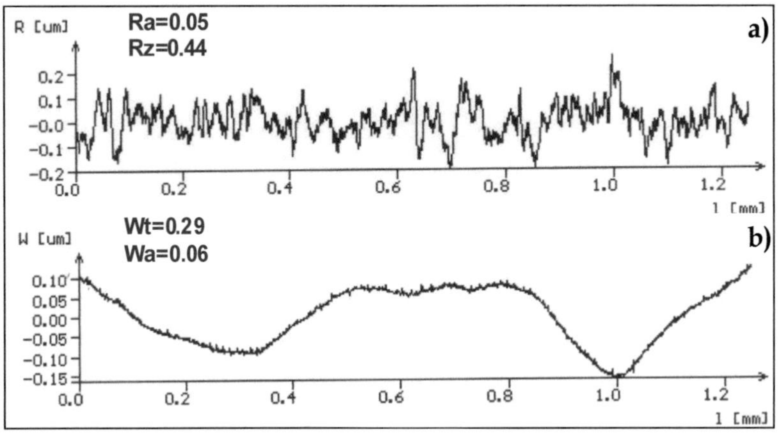


Fig. 12. Profilograms of roughness (a) and waviness (b) in 2D configuration of AZ31 magnesium alloy surface after polishing process

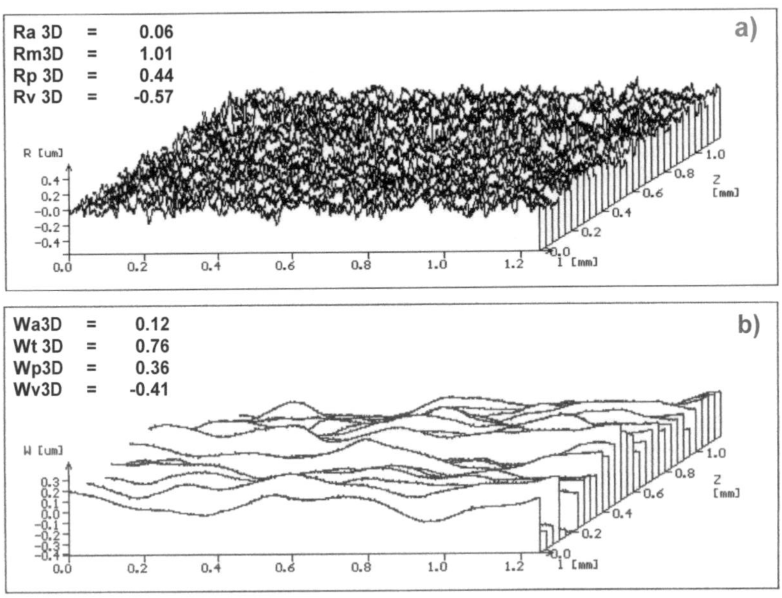


Fig. 13. Profilograms of roughness (a) and waviness (b) in 3D configuration of AZ31 magnesium alloy surface after polishing process

Roughness parameters analysis of investigated samples in 2D configuration (Fig. 12 and 14) has shown insignificant degradation of roughness of samples surfaces with carbon coating manufactured in PACVD process, compared with samples surfaces after polishing process. It has been certified increase of surface roughness, which has carried out accordingly: Ra increase equel to 0,01 μm and Rz increase equel to 0,08 μm. However measurements carried out in 2D configuration have not revealed significant differences in values of surface waviness parameters. Roughness parameters analysis of investigated samples in 3D configuration (Fig. 13 and 15) has shown larger values of Ra, Wa and Wt parameters of investigated samples. Tendency of these changes has been like in 2D configuration

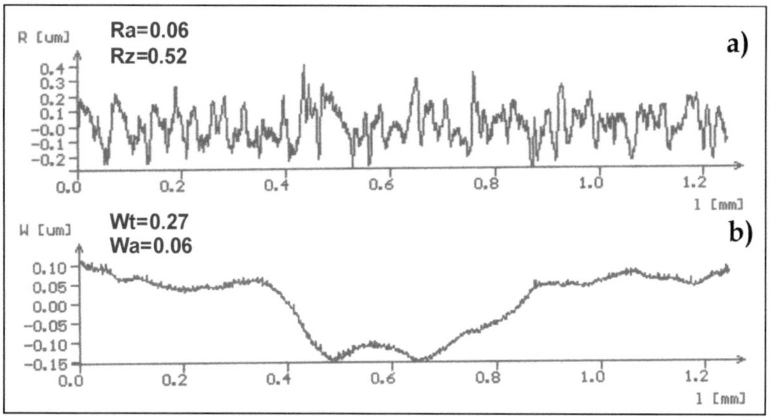


Fig. 14. Profilograms of roughness (a) and waviness (b) in 2D configuration of AZ31 magnesium alloy surface with manufactured carbon coating

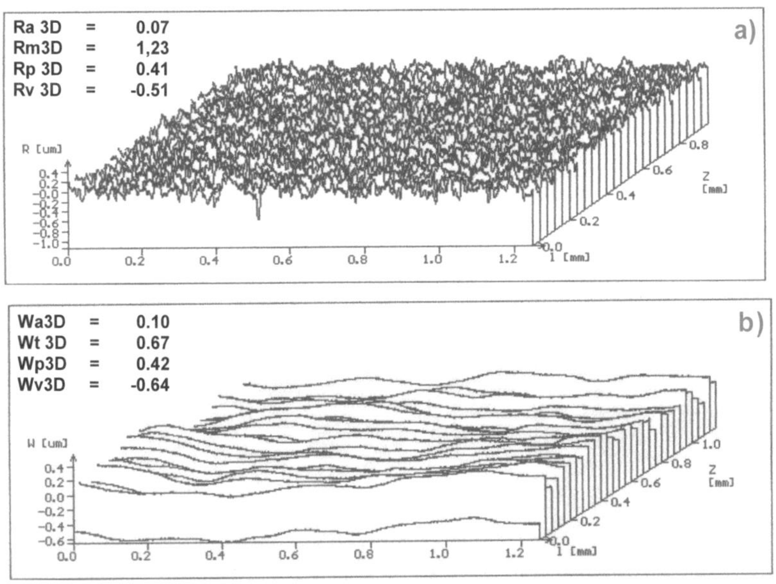


Fig. 15. Profilograms of roughness (a) and waviness (b) in 3D configuration of AZ31 magnesium alloy surface with manufactured carbon coating

measurements. Analysis of profilograms of surface roughness and waviness in 3D configuration has also revealed favourable influence of manufacturing of carbon coating using PACVD method to levelling of maximum profile elevation, expressed by a Ra parameter and maximum profile cavities, expressed by a Rv parameter. Favourable changes observed in above-mentioned measurements have been caused by random deposition of different forms of carbon forms (identified in point No. 7.1) on surface of AZ31 magnesium alloy during PACVD method. Morphology assessment of investigated samples surfaces has concerned comparison of SEM and AFM images of their surface layers, which has been shown in figure 16 and 17. These images have revealed significant changes in their morphology of surface layer. Images of samples surfaces without carbon coatings obtained both by SEM and AFM method (Fig. 16 a and 17a) have shown presence of distinct tool marks caused by polishing process, which have been visible in form of irregular scratches on surface. SEM images of samples surfaces with manufactured carbon coating (Fig. 16 b) have been characterized by mosaic, irregular structure imaging different carbon forms in manufactured coating. Presence of these forma have been confirmed by AFM images. These different forms of carbon in manufactured coatings has had significant influence on their colours. Exemplary AFM image of surface of AZ31 magnesium alloy surface with carbon coating has been presented in figure 17 b.

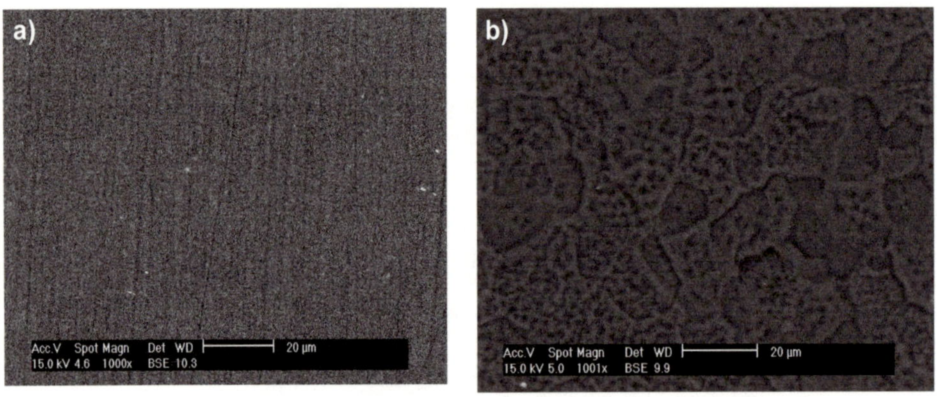


Fig. 16. Images of AZ31 magnesium alloy samples surfaces obtained by scanning electron microscope - SEM: a) surface after polishing, b) surface with carbon coating

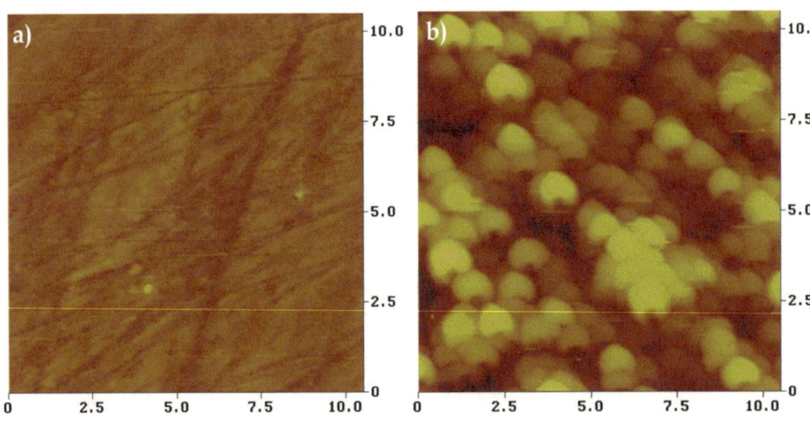


Fig. 17. Images of AZ31 magnesium alloy samples surfaces obtained by atomic force microscope - AFM: a) surface after polishing, b) surface with carbon coating

### 7.5 Tribologic measurements of carbon coatings

Tribologic measurements have relied on tests of rubbing interaction between rectangular specimens of AZ31 magnesium alloy covered by the carbon coating and deprived of this layer with the cylindrical rubbers (Golabczak, 2005). The rubbers have been prepared from three different materials such as hydrogenated rubber butadiene-acrylonitrile - HNBR, poly(methylmetacrylate) - PMMA (plexsiglass), and poly(tetrafluoroethylene) - PTFE (teflon). Dimensions of rectangular specimens of magnesium alloy have been 10x4x5 mm. The cylindrical rubbers has had the diameter of 35 mm and width of 10 mm. Tribologic tests have been conducted using Tribometer T-05 under the following conditions: normal load of the rubber - 6 N, the rubbing speed - 3.67 cm/s, time of test duration - 2 h, frequency of recording of measurements - 2E+14, and ambient temperature (T) of 20.7°C. Representative results of tribologic measurements have been shown in figure 18-19. They have presented differences in the total friction energy and volumetric wear of the examined specimens during the test. The displayed results have provided evidence that the carbon coatings deposited on magnesium alloy have considerably improved their properties. They have both reduced the total energy of friction and enhanced their resistance to wear. The experiments have revealed that the total energy of friction of the listed above specimens with carbon coatings, has been considerably lower and reached: 44% for the rubber one, 130% for the plexiglass rubber and 440% for that made of teflon. Carbon coatings have had also significantly decreased the total volumetric wear of the examined samples of magnesium alloy. The relative increase in wear resistance has been: 660% - in case of the rubber rubber, 540% - for the plexiglass rubber and 800% - for the teflon one.

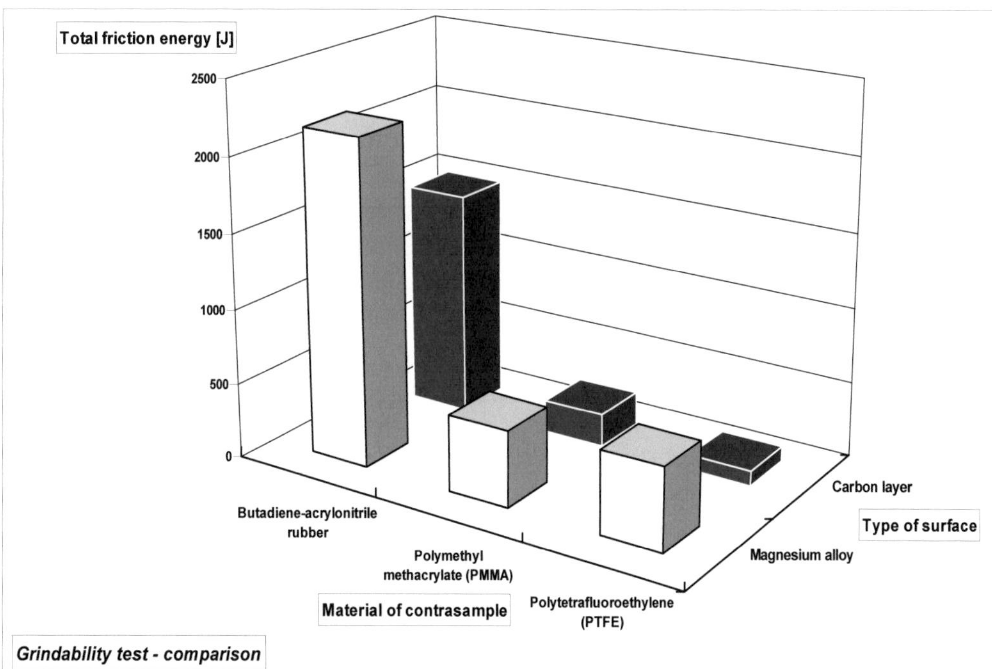


Fig. 18. Comparison of values of the total energy of friction determined by tribologic measurements for specimens made of AZ31 magnesium alloy protected by carbon coating and without the latter

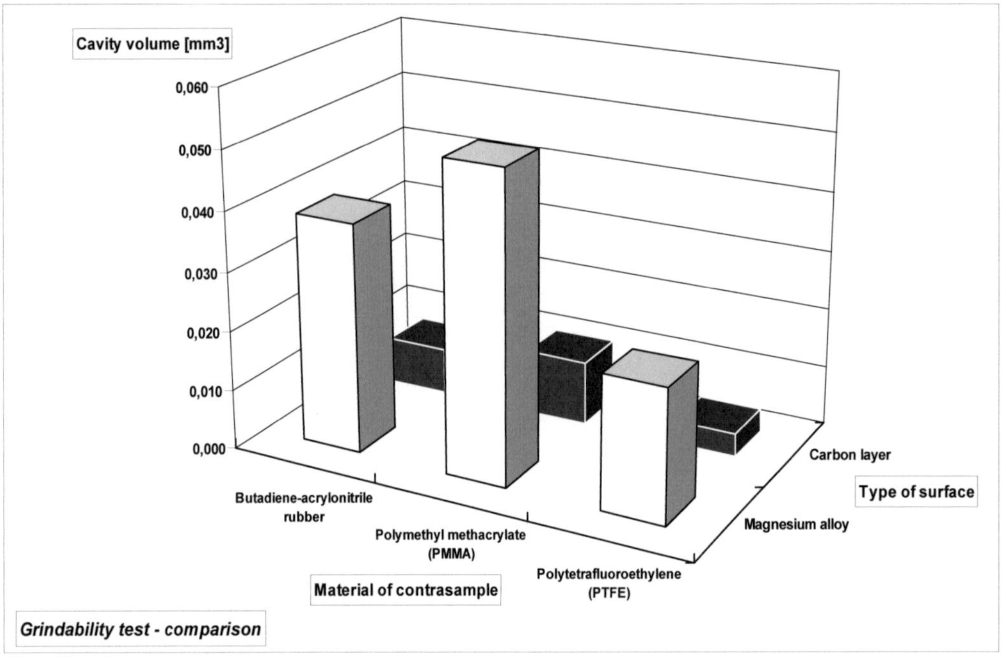


Fig. 19. Comparison of volumetric wear of the specimens made AZ31 of magnesium alloy (one covered by carbon coating and the second - without this coating) determined by tribologic measurements

### 7.6 Determination of corrosion resistance of hard carbon coatings in the salt spray chamber

Corrosion resistance tests of hard carbon coatings have been conducted in SIGMA DIESEL salt chamber (BOSCH). The examined specimens made of AZ31 magnesium alloy have been either protected with the carbon coating or not. Test conditions have been displayed in table 5. The samples without the carbon coating have been exposed to sodium chloride solution for 5 h while the samples protected by this coating have been exposed for 200 h (Golabczak, 2005).

| Test parameters | Value |
|---|---|
| Time of test duration | 5h and 200 h |
| Temperature in the chamber | 35°C ± 1°C |
| Humidity in the chamber | 85% - 90% |
| Intensity of spraying<br>(the sprayed surface area of 80 $cm^2$) | 2ml ± 1ml / h |
| Air pressure | 1.0 bar ± 0.2 bar |
| NaCl concentration in the solution | 5% (w/v) |

Table 5. Parameters of corrosion test carried out in a salt spray chamber

To estimate results of the corrosion test, images of the surface of specimens have been recorded using two microscopes, i.e. metallographic and SEM. Representative images of the surface of examined specimens have been shown in figure 20 and 21.

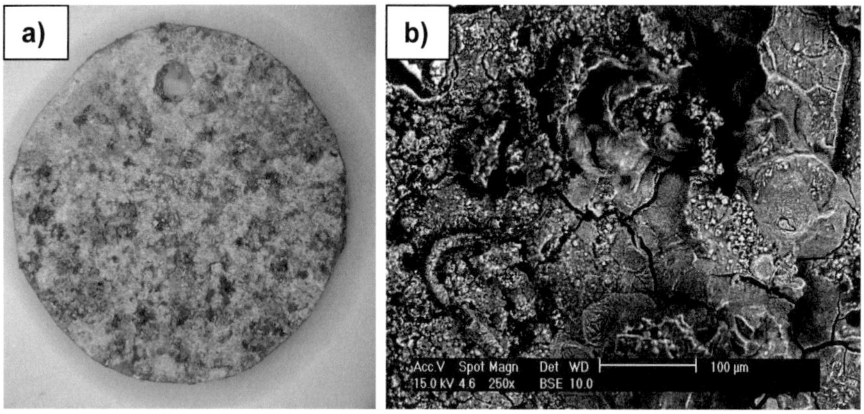


Fig. 20. Microscopic images of AZ31 magnesium alloy samples without carbon coating after 5h exposition in a salt spray chamber: a) magnification 3x, b) magnification 250x

Fig. 21. Microscopic images of AZ31 magnesium alloy specimens with deposited carbon coating after 200h exposition in a salt spray chamber: a) magnification 3x, b) magnification 250x

Comparison of the images of AZ31 magnesium alloy sample surface exposed to the corroding environment in a salt chamber has shown that the samples, which have not been protected by the carbon coating have been strongly corroded after the relatively short time of exposition (5 h). In contrast, the specimens coated by the carbon film have contained only small pits of corrosion after 200 h of treatment under the same conditions. Results of these experiments have demonstrated that carbon coatings explicitly protect magnesium alloy from corrosion.

### 7.7 Determination of corrosion resistance of carbon coatings using electrochemical method

The accelerated electrochemical method has consisted in repeated potentiostatic measurements carried out by using Volta Master 1 set comprising a potentiostat Radiometr-Copenhagen PGP 201. The examined samples have been immersed in Tyrod's electrolyte (its chemical composition is shown in table 6) at the temperature of 20°C.

| $NaCl$ [g/dm$^3$] | $CaCl_2$ [g/dm$^3$] | $KCl$ [g/dm$^3$] | $NaH_2PO_4$ [g/dm$^3$] | $MgCl_26H_2O$ [g/dm$^3$] | $NaHCO$ [g/dm$^3$] | pH |
|---|---|---|---|---|---|---|
| 8.00 | 0.20 | 0.20 | 0.05 | 0.10 | 1.00 | 6.9 |

Table 6. Chemical composition of Tyrod's electrolyte

Modeling of phenomena occurring at the contact interface between the conductor (metal) and the electrolyte has been based on the standard Butler-Volmer equation (Golabczak, 2008). It is a half-empirical equation and characterizes the rate of electric charge transfer through the interface of phases: metal-electrolyte. This rate depends, first of all, on the difference of potentials and its sign (positive or negative) at this interface. The analysis of current flow through the medium which is far from the state of equilibrium cannot be done without the model of Butler-Volmer. The model has based on an electric nonlinear circuit has been proposed to determine the flow of current in the wide range of potential values. This circuit contains some elements responsible for individual physical phenomena that take place during potentiostatic measurements. The scheme of this substitute circuit has been shown in figure 22.

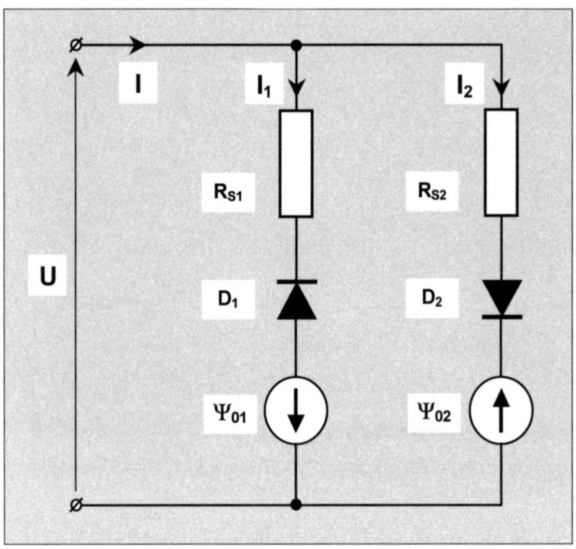


Fig. 22. The structure of proposed model in the form of the nonlinear electric circuit with lumped constants

The proposed model consists of two branches characterizing anodic and cathodic currents. Relationships between the elements of the model shown in figure 22 and phenomena occurring at the interface conductor-electrolyte are as follows:

- diodes $D_1$ and $D_2$ that are fundamental elements of the proposed model correspond to the exponential components of Butler-Volmer equation that are responsible for diffusion ,
- resistors $R_{S1}$ and $R_{S2}$ are responsible for the transfer of electric charge carriers and are particularly important at higher values of voltage U,
- voltage generators $\Psi_{01}$ and $\Psi_{02}$ are responsible for the difference in potentials at the contact interface for anodic and cathodic parts of the characteristics.

Equations describing the substitute electric circuit (Fig. 22) are as follows:
- for the anodic branch (left parts of descending curves in figure 23 and 24):

$$j_1 = -j_{01}\left(e^{-\frac{q \cdot m_1}{k \cdot T} \cdot U_{D1}} - 1\right) \tag{3}$$

$$R_{S1} \cdot j_1 + U_{D1} + \Psi_{01} = U \tag{4}$$

- for the cathodic branch (right parts in ascending curves in figure 11 and 12):

$$j_2 = j_{02}\left(e^{\frac{q \cdot m_2}{k \cdot T} \cdot U_{D2}} - 1\right) \tag{5}$$

$$R_{S2} \cdot j_2 + U_{D2} + \Psi_{02} = U \tag{6}$$

The total current flowing through the interface is the sum of anodic current and cathodic current:

$$j = j_1 + j_2 \tag{7}$$

where: q- the elementary charge of an electron, expressed in [C] [A·s], $q=1.6022 \cdot 10^{-19}$ C;
k- Boltzman constant, expressed in [J/K] [kg·s²/m²·K], $k=1.3807 \cdot 10^{-23}$ J/K;
T- the temperature of the contact interface [K];
$j_{01}$, $j_{02}$ – the density of saturation currents, expressed in [mA/cm²];
$R_{S1}$, $R_{S2}$ – resistances representing the phenomenon of electron transfer, expressed in [Ω];
$m_1$, $m_2$ – coefficients dependent on division of currents and valences of ions in the electrolyte, dimensionless quantities.

Electric parameters of this model for individual potentiostatic curves (Tafel curves) have been identified by means of the least square method and by resolving the system of nonlinear equations by the gradient method. The obtained parameters are effective estimators of the true model parameters. Results of potentiostatic measurements have been plotted in figure 23 and 24. Intercepts of curves presenting the voltage on diodes with OX axis correspond to potentials on the interface metal-solution. The values of electric parameters that have been calculated for the assumed models are presented in table 7 and 8. On completion of potentiostatic measurements the surface of the examined samples has been subjected to SEM analysis. Examples of the recorded images have been shown in figure 25 and 26. Analysis of results of electrochemical studies has revealed that deposition of the carbon coating on the surface of AZ31 magnesium alloy significantly dislocated and increased the difference in potentials (corrosion potential increased) at the interface between the metal and solution (Fig. 24) as compared to the reference sample without the coating (Fig. 23). It provides evidence of the beneficial effect of carbon coating deposited on this alloy because the barrier protecting the latter from electrochemical corrosion has been increased.

Also SEM analysis of the surface of the examined samples has confirmed this desirable impact (Fig. 25 - 26). Only single dark spots (probably very small corrosion pits) have been visible at the surface of the samples coated by carbon coating (Fig. 26) whereas harmful results of electrochemical corrosion have been visible on the whole surface of unprotected AZ31 magnesium alloy (Fig. 25).

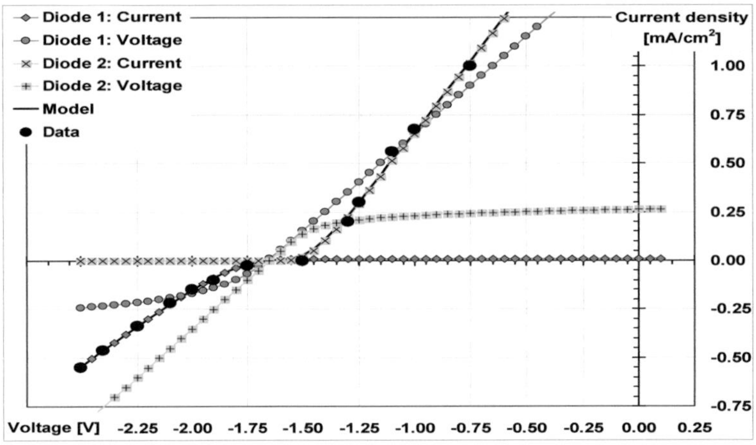


Fig. 23. Potentiostatic curve of electrochemical corrosion for AZ31 magnesium alloy free of carbon coating

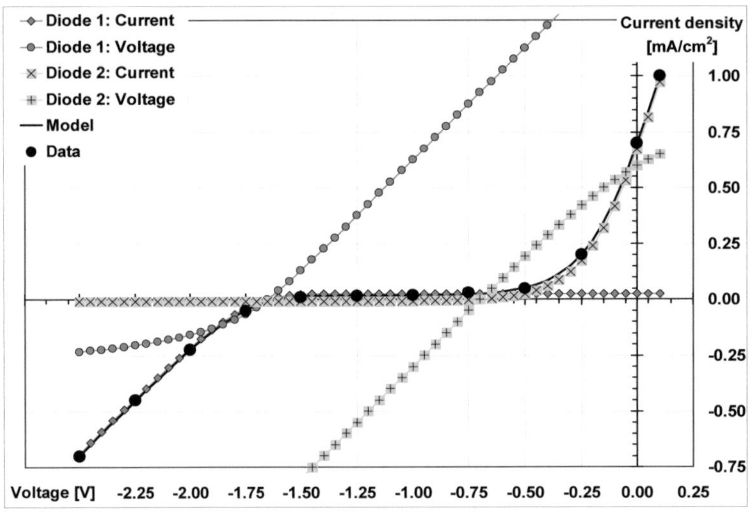


Fig. 24. Potentiostatic curve of electrochemical corrosion for AZ31 magnesium alloy protected by carbon coating

| Model parameters | $J_0$ | $\Psi_0$ | m | $R_S$ |
|---|---|---|---|---|
| | $mA/cm^2$ | V | - | $\Omega$ |
| Diode I | 0.009145 | -1.6622 | 0.4276 | 1092.46 |
| Diode II | 0.000100 | -1.6482 | 0.9607 | 643.96 |
| Fitting error | | 4.027E-03 | | |
| Deviation $3\sigma$ | | 5.280E-02 | | $mA/cm^2$ |

Table 7. Electric parameters of the potentiostatic curve of electrochemical corrosion for sample made of AZ31 magnesium alloy free of carbon coating

| Model parameters | $J_0$ | $\Psi_0$ | m | $R_S$ |
|---|---|---|---|---|
| | $mA/cm^2$ | V | - | $\Omega$ |
| Diode I | 0.026058 | -1.6501 | 0.3588 | 892.26 |
| Diode II | 0.008067 | -0.7000 | 0.1867 | 152.36 |
| Fitting error | | 7.947E-05 | | |
| Deviation $3\sigma$ | | 7.720E-03 | | $mA/cm^2$ |

Table 8. Electric parameters of the potentiostatic curve of electrochemical corrosion for sample made of AZ31 magnesium alloy coated by carbon coating

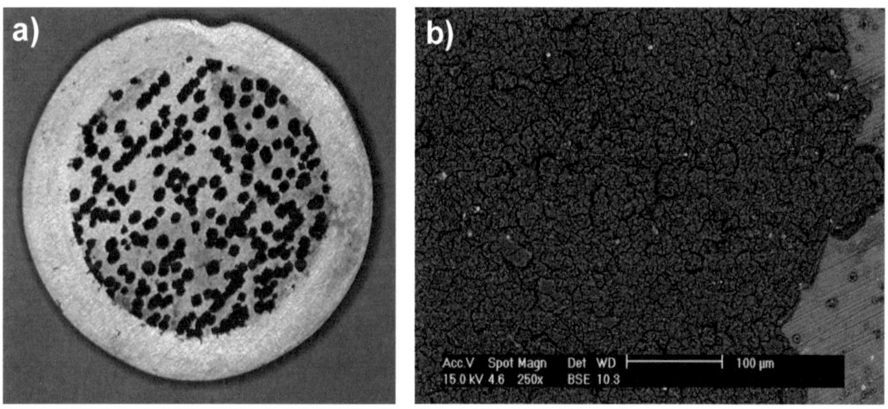


Fig. 25. SEM image of AZ31 magnesium alloy surface on completion of the potentiostatic corrosion test; test duration - 1h: a) magnification 3x, b) magnification 250x

Fig. 26. SEM image of the surface of AZ31 magnesium alloy protected by carbon coating on completion of the potentiostatic corrosion test; test duration 1h: a) magnification 3x, b) magnification 250x

### 7.8 Determination of adhesion of carbon coatings

Adhesion measurement of carbon coating to magnesium alloy substrate has been carried out using scratch tester. The scratch tester is a common method of testing the adhesion of coatings to substrates. For this purpose a special diamond intender is used. The load on the diamond causes stresses to be increased at the interface between the coating and the substrate that can result in delamination of the coating to occur. The load at which the coating first delaminates is called the critical load. For measurement of carbon coating a small diamond intender with a radius of r=200 μm has been applied. During test a diamond intender has been scratched across the coated surface of a substrate at a constant velocity whilst a load has been applied with a constant loading rate F. Applied load F has been changed in the 10 to 30 N range and the load rating has been equal to 5 N. Adhesion assessment of carbon coating to magnesium alloy substrate has been done basing on optical microscope images of wear tracks of worn surfaces of the sample. The critical load of intender Fc, at which occurs cohesion loss has been assumed as criterion of adhesion of the coating to substrate. For ascertainment of this fact the comparison of microscopic images of wear tracks of worn surface of AZ31 magnesium alloy protected by carbon coating has been done (Fig. 27).

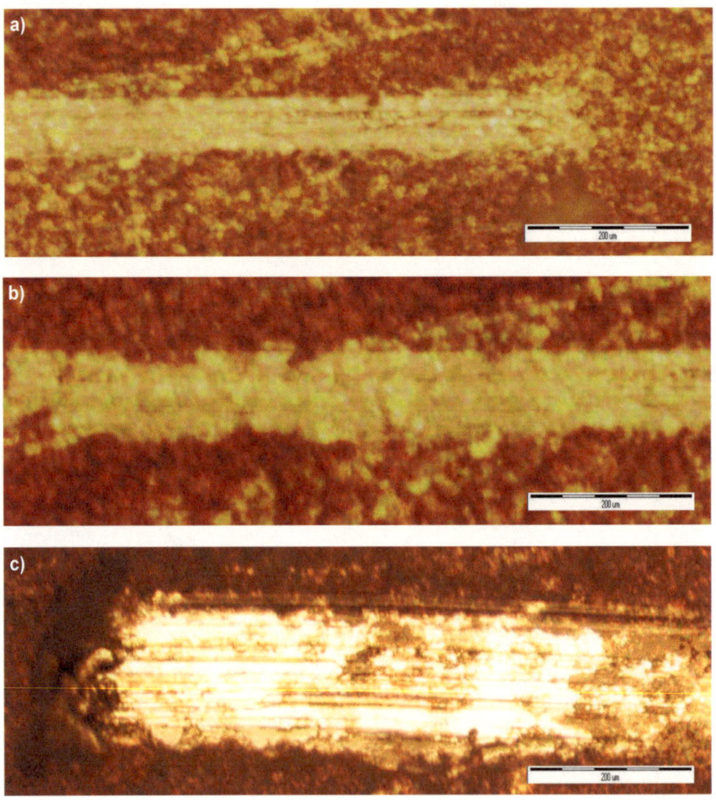


Fig. 27. Characteristic images of wear tracks of worn surface of sample after „scratch test" obtained using optical microscope (magnification 50 x): a) F= 10 N, b) F= 20 N, c) F= 25 N

Presented investigation results have concerned adhesion of carbon coatings manufactured in optimum conditions of PACVD process, i.e. in conditions of test No. 5. Basing on investigations carried out it has been certified that critical load of intender Fc at which occurs no cohesion loss between carbon coating and magnesium alloy substrate has been equal to 20 N (Fig. 27 b). Further increase load of intender (to 25 N) has caused separation of carbon coating from magnesium alloy substrate. This state has been presented in figure 27 c, where sharply outlined surface and white zone of magnesium alloy have been visible.

## 8. Conclusion

Investigations carried out have confirmed that Plasma Activated Chemical Vapour Deposition (PACVD) method enables deposition of protective carbon coatings on AZ31 magnesium alloy. Elaborated technological polishing process of AZ31 magnesium alloy samples has ensured suitable preparation their surfaces, in range of required roughness parameters and their proper purity. Investigations results have confirmed that content of diamond phase in these coatings has been very high. Its relative index $E_i$ has been equal to 0.54 and has depended on PACVD process conditions. Nanohardness of magnesium alloy AZ31 protected by the carbon coating has been considerably higher (24 GPa) than that of the alloy without this coating (0.8 GPa). The measured thickness of a carbon coating has been approximately equel to 220 nm. The displayed results have provided evidence that the carbon coatings deposited on magnesium alloys have considerably improved their properties. They have both reduced the total energy of friction and enhanced their resistance to wear. The experiments have revealed that the total energy of friction of the listed above specimens with carbon coatings, has been considerably lower and reached: 44% for the rubber one, 130% for the plexiglass rubber and 440% for that made of teflon. Carbon coatings have also significantly decreased the total volumetric wear of the examined samples of magnesium alloy. The relative increase in wear resistance has been: 660% - in case of the 540% - for the plexiglass and 800% - for the teflon one. Experiments have demonstrated that carbon coatings explicitly protect magnesium alloy from "salt spray" corrosion and electrochemical corrosion. Basing on investigations carried out it has been certified that critical load of intender Fc at which occurs no cohesion loss between carbon coating and magnesium alloy substrate has been equel to 20 N. Manufactured carbon coatings apart from very attractive operational properties mentioned above, have had also decorative properties. They have had gold glossy colour what can be useful in making of jewellery. Moreover, manufactured carbon coatings have been characterized by very good biocompatibility (because of high content of pure diamond). This property is especially important in protection of implants surfaces. Presented results have indicated that the studies should be continued in order to improve PACVD method and produce carbon coatings on specimens with more complex shapes and on internal surfaces of specimens made of magnesium alloys.

## 9. Acknowledgement

This work has been partially financed by the Grant No. 3 T08C 06826 from Polish Committee for Scientific Research.

## 10. References

Clapa, M.; Mitura, S; Niedzielski, P; Karczemska, A. & Hassard, J. (2001). *Diamond and Related Materials*, Vol. 10, No. 3, pp. 1121-1124

Golabczak M. (2005). *Manufacturing of Carbon Coatings on Magnesium Alloys by PACVD Method*, Technical University of Lodz, Poland.

Golabczak, M. (2007). *Advances in Manufacturing Science and Technology*, Vol. 31, No. 2, pp. 81-90 ISSN 1895-9881

Golabczak, M. (2008). *Mechanics and Mechanical Engineering*, Vol. 12, No 2, pp. 157-164, ISSN 1428-1511

Golabczak, M. & Konstantynowicz, A. (2009). *Journal of Achievements in Materials and Manufacturing Engineering*, Vol. 37, pp. 270-276

Golabczak, M. & Konstantynowicz, A. (2010). *Defect and Diffusion Forum*, Vols. 297-301, pp. 641-649, ISSN 1012-0386

Golabczak, M. (2010). *Advanced Knowledge Application in Practice*, Sciyo, ISBN 978-953-307-141-1, Rijeka, Croatia

Gray, J.E. & Luan, B. (2002). *Journal of Alloys and Compounds*, Vol. 336, No 1, pp. 88-113, ISSN 0925-8388

Hawkins, J.H. (1993). *Global View Magnesium: Yesterday, Today, Tomorrow*, International Magnesium Association, pp. 46-54

Ida, T.; Ando M. & Toraya H. (2000). *Journal of Applied Crystallography*, Vol. 33, pp. 1311-1316

Ishizaki, T.; Shigematsu, I. & Saito, N. (2009). *Surface & Coatings Technology*, Vol. 203., No. 16, pp. 2288-2291, ISSN 0257-8972

Kawalla, R.; Lehmann, M. & Ullmann, M. *Archives of Civil and Mechanical Engineering*, Vol. VIII, No. 2, pp. 93-101, ISSN 1644-9665

Kuc, D.; Hadasik, E.; Niewielski, G. & Plachta, A. (2008). *Journal of Achievements in Materials and Manufacturing Engineering*, Vol. 27, pp. 27-30

Niedzielski, P.; Mitura, E.; Mitura, S. et al. (1997). *Diamond and Related Materials*, Vol. 6, No. 5, pp. 721-724

Olszyna, A. & Smolik, J. (2004). *Thin Solid Films*, Vol. 459, No 1/2, pp. 224-227

Robertson, J. (2002). *Materials Science and Engineering*, Vol. 37, 2002, pp. 129-281

http://bbs.com (2008)

http://buehler.com (2010)

http://olympus.com (2011)

# 4

# Fatigue Cracking Behaviors and Influence Factors of Cast Magnesium Alloys

Xi-Shu Wang
*Department of Engineering Mechanics, School of Aerospace, AML, Tsinghua University, Beijing, China*

## 1. Introduction

Accurate prediction of fatigue performance and life is a key issue in the design and applications of the high strength-to-weight ratio of many contemporary magnesium alloys, which are suitable for ultimate weight reduction purposes in automotive and aircraft components. For example, the ever increasing demands for higher efficiency and light weight typical cases of power generation, aerospace and automotive industries call the maximum exploitation of the material's properties. The combination of low density (about 1.74 g/cm$^3$), high specific strength and excellent castability qualifies magnesium alloys as ideal materials for the lightweight constructions [1]. Thus, purely empirical models that heavily rely on larger safety factors are of limited uses. It is now widely recognized that fatigue damage models that are closely related to the microstructural features to provide a more reliable basis for a life prediction, provided that the relevant microstructural damage mechanisms are accurately accounted for. Therefore, there are increasing interests in the fatigue tests combined with the high-resolution microscopic techniques, environmental influencing factors, and especially in understanding their fatigue crack initiation and propagation behaviors with process rationalization [2-4]. Subsequently experimental observations on cast magnesium alloys have accumulatively revealed that the dendrite cell size, pores, secondary phase particles, persistent slip bands and twinning in the dendrite cells considerably affect on the fatigue durability and fatigue cracking behavior or crack growth mechanism of dendritic magnesium alloys [5-8]. The quantitative estimation of small fatigue crack growth rate of cast magnesium alloys is whether simply and effectively to assist the reversed design of cast magnesium alloys or ceaselessly to improve the strength and toughness of cast magnesium alloys. At the same time, it is necessary how to understand that these microstructural features of cast magnesium alloys play role in the fatigue cracking behavior or that the microstructural evolution reacts to the applied loadings. However, it is difficult to use conventional alloying techniques to improve some of their properties, e.g. elastic modulus, elastic-plastic deforming property and the different thermal expansion between phases during the shrinking at the high elevated temperatures. Under these conditions the elastic, and the possible plastic, properties of the secondary phase will influence the mechanical response during an applied loading since the interface of secondary phase will transmit stresses from the matrix around the secondary phase when the interface both the matrix and secondary phase has a compatible strain field. If the elastic modulus or elastic-plastic property of the secondary phase is larger than that of the surrounding matrix in the secondary phase will take up a larger load than the matrix next to the secondary phase so that the stresses are reduced in the vicinity of the secondary phase. On the other hand, the modulus of elasticity of the secondary phase is smaller that that of the matrix, the secondary phase will take up a smaller load than the matrix next to the secondary phase so that the stresses are increased in the vicinity of the secondary phase. From the above analysis it is obvious that the secondary phase with different thermal expansion properties or elastic properties compares to the matrix will behave differently with respect to crack initiation and early growth during fatigue loading. Therefore, one must resort to fiber/particle reinforcement in order to reduce the difference of thermal expansion properties or elastic properties between the secondary phase and the matrix. The solubility of alloying elements in magnesium alloy is limited, which restricts the possibility of improving the mechanical properties and chemical behavior of this material. The crystal structure of magnesium is hexagonal which limits its inherent ductility. The only alloying element, which causes a useful phase change to bcc, in this respect, is lithium. The property profiles demanded by automobile and other large-scale potential users of magnesium alloys have revealed the need for alloy development. A direct transfer of high performance aircraft alloys is not possible on economic grounds and the property profile does not coincide. Ebert et al. [1] previously indicated that there are four development trends based on their main requirements, which are as following respectively: First trend of specific strength is Mg-Al-Mn, Mg-Al-Zn, Mg-Zn-Cu, Mg-Al-Ca (-Re), Mg-Li-X. Second trend of good ductility is Mg-Si, Mg-Al-Ca (-re), Mg-Li-X. Third trend of good creep resistance is Mg-Al-Re, Mg-Al-Ca-X, Mg-Ag-Re-Zr, Mg-Y-Re-Zr, and Mg-Sc-X-Y. And the final trend of good wear, creep, thermal expansion is fibre/particle reinforced Mg-MMC's. Therefore, the mechanical properties and microstructure of magnesium alloys can be improved by above mentioned development trends [1,9-11]. No matter what development trends in the alloying processes of this material, some microstructural defects are impossibly to be avoided. And, the effects of these microstructural defects on the cracking behavior, the interactions on the microstructural defects and differential phases with different elastic or thermal expansion properties occur mainly in the ranges of meso/micro scales; therefore, the investigations have to localize in the techniques with high-resolution microscope, especially *in-situ* measurements.

Dating back to the seventies of the last century, the potential of *in-situ* fatigue studies conducted in a scanning electron microscope (SEM) has been realized as SEMs combine the servo-hydraulic loading (provided thermal loading, humid environment etc.) testing system. Since that time, numerous studies have demonstrated that *in-situ* fatigue in the SEM device is an effective tool for the investigation of microstructural effects on the development of slip features [12,13], on crack initiation [2,5,14-19], for the analysis of the mechanisms of crack growth [14-17,20,21], for the quantitative determination of small crack growth rates [14-18,22], crack tip opening displacements [16,23-25], for the examination of interactions between fatigue crack growth and microstructure [15,16,26-28], for the effects of vacuum, high temperature, water vapor environments, notch and pores alignment etc. [6,15,16,29-31]. For example, the majority limitation of most studies were imposed by the fact that the experiments had to be conducted at room temperature (RT) in a high-vacuum, whereas most components typically operate in a gaseous environment, which can simultaneously be combined with the high-temperature conditions and after prior-corrosion magnesium alloys. This is because the magnesium alloys are easily corroded and to have the low creep resistance. With the advent of environmental scanning electron microscopes (ESEM), the situation has changed considerably. Today's ESEMs typically allow for imaging in various gaseous environments at a pressure up to about 20-30 Torr, and thus, high-resolution data may be obtained under the conditions relevant to those of actual components. Most procedures required for *in-situ* fatigue testing conducted at RT in an ESEM are not much different from those required for tests in high vacuum. Environmental *in-situ* fatigue testing at elevated temperatures and in corrosion conditions, especially in micro/meso scales, is still in its beginning, and it is hoped that the current study will stimulate a further research in this area, which includes to the multi-scales and multi-dimensions fracture/fatigue damage mechanics in the reversed design of magnesium alloys with some processing defects.

Following discussions and analyses about the fatigue cracking behaviors and influencing factors of typical cast magnesium alloys are the main aims of this chapter. These involve the considerations of materials science, engineering mechanics and experimental technology at the microstructural level including to the microstructural defects. Principles of these sciences and technology are utilized for a design of microstructures having superior fracture/failure resistance. The fatigue micro crack initiation and propagation behaviors are especially useful the design against fracture, in providing the improving the approach about the mechanical properties of cast magnesium alloys. The design philosophy inherent to fracture mechanics is that the operating stress must be less than the magnesium alloys' critical fracture stress. The critical stress may be much less in "real" engineering structures than in "flaw-free" magnesium alloy as a result of notches, cracks and pores of macroscopic dimensions that are present to one degree or another in all structural parts. Therefore, the room temperature tests conducted on the cast magnesium alloys are addressed first, as these demonstrate that substantial environmental effects on the fatigue damage evolution can be observed in certain materials even at room temperature. The second part of the chapter focuses on high-temperatures *in-situ* testing, and the provided examples are from recent fatigue tests on the results of fatigue cracking behavior and influencing factors for the typical cast magnesium alloys, which involve the effects of spacing and alignment with two pores on the fatigue crack initiation and propagation behaviors etc.

## 2. Phenomenological descriptions of typical cast magnesium alloys

### 2.1 Microstructural characteristics

Scanning electron micrographs at the lower and higher magnifications of typical cast magnesium alloys, such as AZ91D, AZ91, AM60, AM60B, AM50 and additions some mischmetals into these magnesium alloys, were shown in as following Figures. In these SEM images, we see clearly the mainly microstructural components of primary α-Mg cells and β-$Mg_{17}Al_{12}$ phase as shown in Figs. 1A-1E. With increasing of aluminum content, the distribution density of β-$Mg_{17}Al_{12}$ phase becomes larger and larger. As the typical cast AM50 alloy, the microstructural components should consist of the following:

1. Hexagonal α-Mg cells having a linear intercept average size of about 20 μm. However, larger α-Mg cells with an average size of about 50 μm appear in groups within isolated nodules (Figs. 1E). The microstructure of other cast magnesium alloys such as AM60, AZ91, has a similar to that of AM50. The obvious difference of their microstructures is at the size and distribution of α-Mg phase or in the geometrical shape of β-$Mg_{17}Al_{12}$ phase. For example, microstructural morphology of AM60 or AZ91 alloy consists also of α-Mg matrix and secondary phase of β-$Mg_{17}Al_{12}$ or β-$Mg_{17}Al_{12}X$ (X denotes another

element) along grain boundaries as shown in Figs. 1B and 1C, especially in thick section with lower cooling rate.

2. In these microstructures, there is a little other Mn rich phase inclusion juxtaposed of two particles (as shown in Figs. 1D). As illustrated result in Figs. 1D, this small white particle on the secondary phase represents a distinct phase in a polygonal shape more enriched in Mn and centered on an oval grey precipitate. The polygonal particle is reminiscent of the hexagonal crystal structured $Al_8Mn_5$ phase reported by Wang et al. in their transmission electron microscopy (TEM) analysis [32]. From the electron probe microanalysis mapping, this phase still contains Mn element but with more enrichment in Al element and a maximum localization of Si element. It may correspond then to an agglomeration of $Al_6Mn$ or/and AlMnSi phase reported in Refs [33] and [34], respectively. I agree much more that the aforementioned superposition of Mn-rich particles is referred to as Mn-rich inclusion [4].

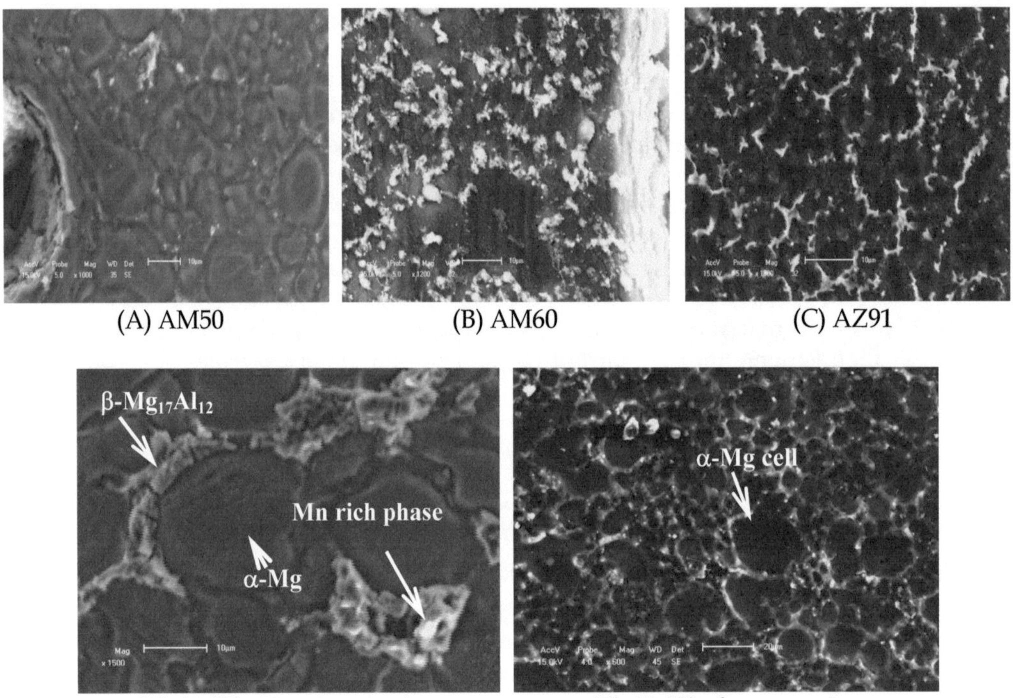


(A) AM50 (B) AM60 (C) AZ91

(D) AM50 prior to corrosion (E) AM50 after corrosion

Fig. 1. Typical microstructure of cast magnesium alloys

3. In addition, the shrinkage pores or other casting defects are also the microstructural feature of cast magnesium alloys as shown in Figs. 2. The shrinkage pores or casting defects are caused by the different deformation rates because there are different thermal expansions and elastic properties between the two phases. These defects of shrinkage clusters located disjointedly in the interdendritic regions. These pores cause the hydrogen penetration into the dendritic structure during solidification, in which the surface of pores occurs easily in the significant plastic striations in micro scale under the applied loading [4].

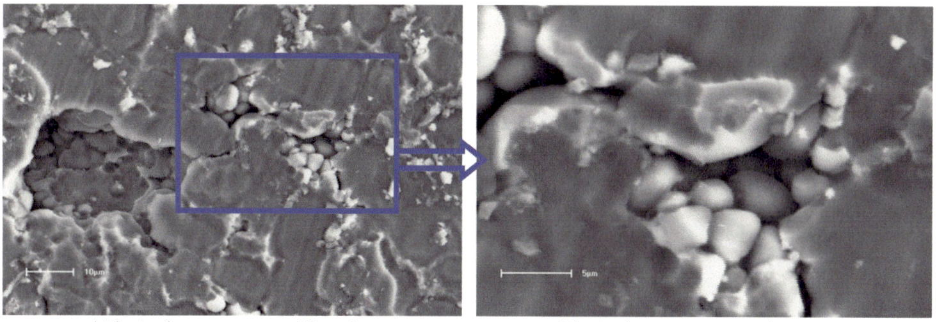

(A) At lower magnification (B) At larger magnification

Fig. 2. Typical microstructural defect of cast AM50 alloy

4. The microstructure of typical cast AM50 alloy is present in the brittle eutectic phase (β-$Mg_{17}Al_{12}$) that its micro-hardness is about 88.34 MPa, and is present the softer matrix of primary phase (α-Mg) that its micro-hardness is about 66.86 MPa [15]. As the existent difference of micro-hardness between the two phases, the interface of two phases occurs easily in the deformation mismatch under the applied loading [4,6,10,14-17]. It is thus clear that a pore which is debonded from the matrix will act as a preferential site for crack initiation and crack early growth. The only significant difference between a surface and an internal pore/debonded β-$Mg_{17}Al_{12}$ is that the former will show more rapid crack growth once the crack has circumvented around the pore/debonded β-$Mg_{17}Al_{12}$.

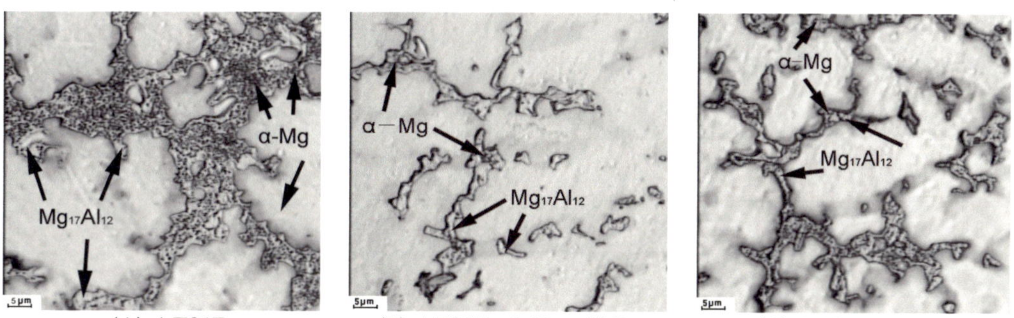


(A) AZ91D (B) AZ91D+0.40 wt%Ca (C) AZ91D+0.40 wt%Ca+0.05 wt%Sr

Fig. 3. Refinement microstructure of cast magnesium alloys with addition of Ca/Sr

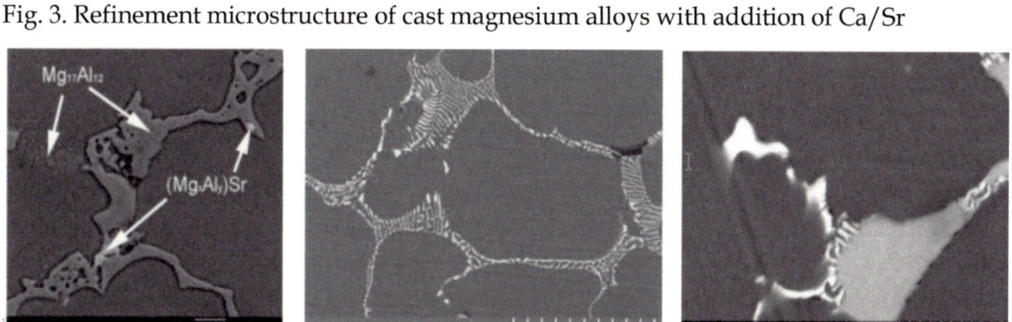


(A) AZ91D + 0.10 wt%Sr (B) AM50+ 1.20 wt%Sr (C) AM50+1.80 wt%Sr

Figs. 4. ($Mg_xAl_y$)Sr phase distribution of cast magnesium alloys with addition of Sr

5. The rare earths (REs) or Ca etc. have been used in magnesium alloys for many years, whereas the alloys of Mg-Al-RE/Ca/Sr system have been developed in the recent decade [35,36]. For example, the refinement influence of the separated and composite addition of Ca and Sr is showed in Figs. 3. We can see that the microstructural features of cast magnesium alloys after refinement influence based on the addition some elements into the matrix. One of many influencing factors is that the β-phase becomes more and more slender when the addition of 0.4 wt%Ca into AZ91D was carried out as shown in Figs. 3A-3B. And other phase such as $(Mg_xAl_y)Sr$ located mainly on the β-$Mg_{17}Al_{12}$ phase thus it can classify as the secondary phase as shown in Figs. 4. Through the addition RE or Ca/Sr into the Mg alloys, formed the α-Mg grain boundaries can cause the interface to become a more steady than that prior to addition. The behavior of addition REs or Ca/Sr can improve the mechanical, corrode and hot-crack resistance properties, especially the mechanical property of cast magnesium alloys at high temperatures [9-11,37,38].

## 2.2 Mechanical properties

As above mentioned, some microstructural features, such as a dendrite, interdendritic pore, shrinkage pores, and secondary phase particles, are the important factors to affect on the mechanical properties. On the other hand, the boundary strengthening by addition RE or Ca/Sr elements can improve the mechanical properties of cast magnesium alloys. Therefore, the main chemical compositions and mechanical properties of typical cast magnesium alloys used the fatigue cracking tests in this chapter were listed in Table 1. And in these typical cases, the effects of addition Ca/Sr into AZ91D alloy on the mechanical properties were listed as shown in Table 2.

| **Alloys** | **Al** | **Mn** | **Zn** | **Si** | **Yield stress** $\sigma_{0.2}$ **(MPa)** | **Tensile stress** $\sigma_b$ **(MPa)** | **Elongation δ%** |
|---|---|---|---|---|---|---|---|
| AM50 | 4.5-5.3 | 0.2-0.5 | <0.2 | 1.2 | 135 | 200 | 9-10 |
| AM60 | 5.99 | 0.2 | 0.22 | 1.2 | 160 | 210 | 4-7 |
| AM60B | 5.5-6.5 | 0.2-0.6 | 0.25 | 1.0 | 150 | 240 | 10-12 |
| AZ91 | 8.2-9.5 | 0.15-0.4 | 0.92 | 0.05 | 180 | 250 | 3.0-3.5 |
| AZ91D | 8.3-9.7 | 0.15-0.5 | 0.35-1.0 | 0.10 | 72 | 160 | 1.7-1.9 |

Table 1. Main chemical compositions and mechanical properties of typical cast Mg alloys

In Table 2, it is clearly seen that the yield stress of AZ91D increase slightly but the tensile stress and elongation decreases slightly with the increasing addition Ca amount into the cast AZ91D at RT. The mechanical properties improved well under the slightly composite addition of Ca and Sr. It is because that ductility parameter as mechanical properties of cast magnesium alloy is determined by the number of operative slip systems. Mg being hexagonal slips at room temperature on the base plane (0001) <1120> and secondary slip on vertical face planes (1010) in the <1120> direction. This limits ductility at the lower temperatures. At elevated temperatures slip also occurs in the $<11\hat{2}0>$ direction on the

$(10\bar{1}1)$ pyramidal planes. This behavior is influenced by an alloying, but as long as the structure remains hexagonal the effects based on Mg-Si and Mg-Al-Ca-(RE) and Mg-Li-X,

| **Composition** (Wt %) | **RT tensile** | | | **150 °C tensile** | | |
|---|---|---|---|---|---|---|
| | Yield stress $\sigma_{0.2}$ (MPa) | Tensile stress $\sigma_b$ (MPa) | Elongation $\delta$ % | Yield stress $\sigma_{0.2}$ (MPa) | Tensile stress $\sigma_b$ (MPa) | Elongation $\delta$ % |
| AZ91D | 72 | 160 | 1.67 | 58 | 105 | 4.8 |
| AZ91D-0.40Ca | 83 | 145 | 1.30 | 70 | 97 | 3.0 |
| AZ91D-0.60Ca | 84 | 143 | 1.30 | - | - | - |
| AZ91D-1.00Ca | 86 | 138 | 0.99 | - | - | - |
| AZ91D-0.40Ca-0.05Sr | 80 | 152 | 1.49 | 68 | 112 | 3.7 |
| AZ91D-0.40Ca-0.10Sr | 76 | 150 | 1.51 | 65 | 108 | 3.9 |

Table 2. Mechanical properties of the typical cast Mg alloys with an addition Ca/Sr

which offer the possibility of having a phase mixture of bcc and hcp phase, there has been punch work into the development of fine grain material. One such technique, which has been proved successful in the case of aluminum and copper based alloys and steels, is spray forming fine grain material [1]. However, the composite addition a little Ca/Sr into AZ91D caused the yield stress of cast magnesium alloy much more to be improved than that single addition Ca into AZ91D. Another importance mechanical property of cast magnesium alloy is the tensile behavior (including to the tensile strength and ductile properties of material), which comes from the typical tensile tests about the cast AM60B and AZ91D at room and elevated temperatures were carried out with the smooth specimens. The tensile behaviors by these tensile curves of cast magnesium alloys as shown in Figs. 5 are better than that by Table 2. In the Figs. 5A, the effect of the elevated temperature (which is over than 150 °C or 423K) on the strength of cast AM60B can be not ignored but the effect of the elevated temperature on the strength of cast AM60B+0.05wt%Sr is not obvious as shown in Figs. 5A. This is because the addition slight Sr into AM60B caused the α-Mg grain coarsening [38]. The coarsening grain will influence the tensile properties of AM60B compared with the results before and after addition of slight Sr amount at the RT condition but with a slight influence on the tensile properties at the elevated temperature. However, for the cast AZ91D, before and after addition 0.05-0.10 wt%Sr, you can clearly see that the effect of elevated temperature and addition a slight Sr amount on the strength and ductile properties can be also not ignored as shown in Figs. 5B. But, there is still not clear issue on the influence mechanism of cast magnesium alloys nowadays during these additional processes and at the critical transferred temperature of cast magnesium alloys. Many researchers thought that mischmetal has grain refinement effect on AZ31, AZ61 and AZ91 alloys, and put forward the grain refinement mechanism as follows. During the solidification process, mischmetal enriched in front of solid/liquid interface, and gave rise to growth of α-Mg dendrites was suppressed and resulted in grain refinement. But no micrographs were given showing grain structures before and after mischmetal addition into AZ31, AZ61 and AZ91 alloys [38]. Another literature mentioned that small addition of pure cerium and mischmetal consisting of La and Ce led to considerable grain coarsening in AM50 alloy, and grain structures before and after mischmetal addition were shown, however, there was no explanation about the grain coarsening mechanism [39]. Although the experimental conditions in each researcher were different, the influence of mischmetal addition on grain size of Mg-Al alloys should be the same. The main reason for the discrepancy ascribed to the difficultly etching technique for revealing the grain boundaries of casting magnesium alloys. Li et al. [38] reported that small amount addition of mischmetal ranging from 0.10 wt% to 1.20 wt% (mass fraction) causes considerable grain coarsening in Mg-Al alloys. After adding 1.20 wt% (mass fraction) mischmetal into AZ31, AM60 and AZ91 alloys respectively, the average grain size is approximately twice as larger as that of the original alloy. The change of microstructure must cause the change of mechanical properties of cast Mg-Al alloys.

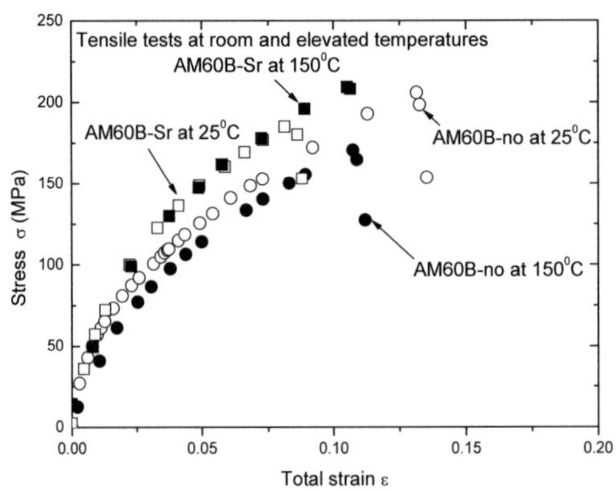


(A) Stress versus strain of AM60B

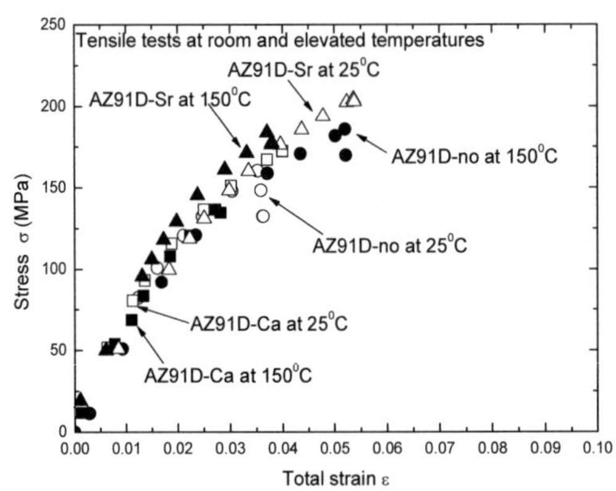


(B) Stress versus strain of AZ91D

Fig. 5. The tensile behavior of typical cast magnesium alloys

We previously reported that with increasing the amount of addition Ca into AZ91D, the hot-crack property decreases but the α-Mg grain-refinement of AZ91D alloy was improved [9-11]. The hot-cracking mechanism of Ca addition into AZ91D alloy was conjectured to be that Ca addition elevates the tendency of the divorce eutectic and formation temperature of the

$Al_2Ca$ phase. The Ca-contained phase was distributed as a net-shape on the grain boundary and debases the boundary tension of liquid film, deteriorating the filling capacity and lowering the hot-crack property of cast magnesium alloys [9,11]. The influence of addition Sr on the microstructure of Ca-containing AZ91D alloy is shown in Figs. 3C. The Sr element aggregates also to the grain boundary just as Ca element does. Some of Ca element combined with Sr and formed a certain compound phase labeled MgAlCaSr phase. Sr addition to Ca-containing AZ91D alloy effectively suppresses the influence Ca addition on AZ91D alloy, which can assist in improving hot-crack resistance of Ca-containing AZ91D alloy. The effects of Ca/Sr separate and composite additions into cast AZ91D magnesium alloy on the microstructure and tensile behavior indicated that Ca refines both grain and eutectic phase of AZ91D magnesium alloy. However, Sr can weaken the refinement effect of Ca when Ca/Sr composite is added. Ca/Sr addition evidently improves the microstructural stability of AZ91D magnesium alloy at elevated temperature. Addition of small Ca amount into AZ91D magnesium alloy improves yield strength but decreases the elongation of this alloy. Appropriate Ca/Sr composite addition into AZ91D magnesium alloy can improve the elongation and maintain the excellent yield strength of AZ91D magnesium alloy containing Ca. It is deduced that separate Ca addition to AZ91D magnesium alloy is able to refine grain size and consequently improve the yield strength, based on the Hall-Petch relation, which is shown in Fig. 6. With decreasing the α-Mg grain diameter ( $d$ ), the yield stress increased at the range from 170 μm ( $\frac{1}{\sqrt{d}} = 0.077$ ) to 110 μm ( $\frac{1}{\sqrt{d}} = 0.095$ ) in the different temperatures as shown in Fig. 6. The differences in segregation of Ca and Sr in a diffusion layer ahead of the advancing solid/liquid interface during solidification affects the process of grain growth, which may result in grain size variation. Ca possesses strong segregation ability in the melt alloy and can promote an intensive constitutional under cooling during cooling in a diffusion layer ahead of the advancing solid/liquid interface that restricts grain growth [10].

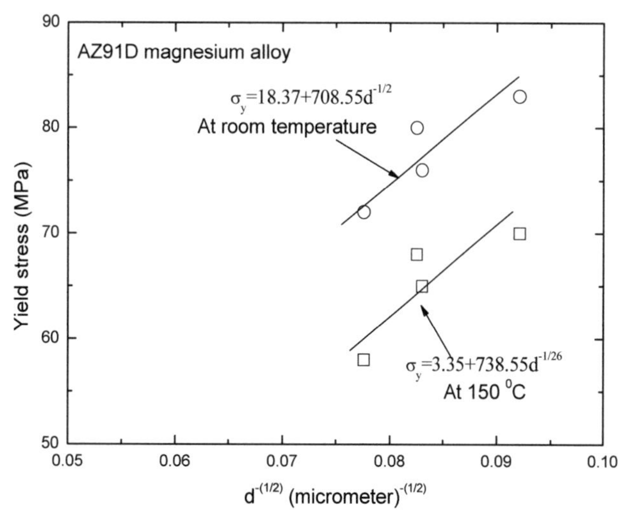


Fig. 6. Yield stress at different temperatures versus the diameter of $d$

### 2.3 Low-cycle fatigue cracking mechanism

#### 2.3.1 Fatigue cracking mechanism at room temperature

Based on the above mentioned microstructural features of typical cast magnesium alloys, including addition Ca/Sr or RE elements into these alloys, the low-cycle fatigue (LCF) crack initiation and propagation tests were carried out at RT by using scanning electron microscopy. Majority experiment results indicated that the fatigue crack initiation occurred preferentially at the pores, grain boundary because of the slip incompatibilities between adjacent dendrite cells as shown in Figs.7 [6,12,13-17]. In the Figs. 7A, the fatigue crack initiation occurred at the root of manual notch which its radius is about 50 μm and it is slightly larger than the average size of α-Mg grain of cast AM50 alloy. The micro crack propagated in interdendritic regions, along crystallographic planes. The propagation behavior of small fatigue crack depended strongly on the effect of both high temperature and microstructure. When there is a pore as shown in Figs. 7B, the crack occurred at the pore and accompanied by slight plastic deformations in the adjacent α-Mg grain surface. Another aspect of cracking shows in Figs. 7C, which shows that the crack initiation mechanism of general cast magnesium alloys is due to the plastic deformation incompatibilities to occur at the boundary or pores between the adjacent grains. Even if there are not the slip vestiges on the surface α-Mg grains under the applied stress, the crack occurred also at the boundary because the deformation mismatch between the α-Mg phase and β-$Mg_{17}Al_{12}$ phase as shown in Figs. 7D-7F. One of important reasons is caused by the difference of micro-hardness between both two phases [6,13-16]. As this reason, the fatigue crack propagated mainly along the boundary or β-$Mg_{17}Al_{12}$ phase of cast magnesium alloys at the RT as the typical cases shown in Figs. 8. The fatigue crack propagation process of cast typical AM50 alloy indicated that the crack propagated microscopically in the zigzag manner but macroscopically in a projection to the loading direction. With increasing of cyclic number under the maximum applied stress of 128 MPa ($\sigma_{max} / \sigma_{0.2} = 0.95$) at $R$=0.1, the fatigue

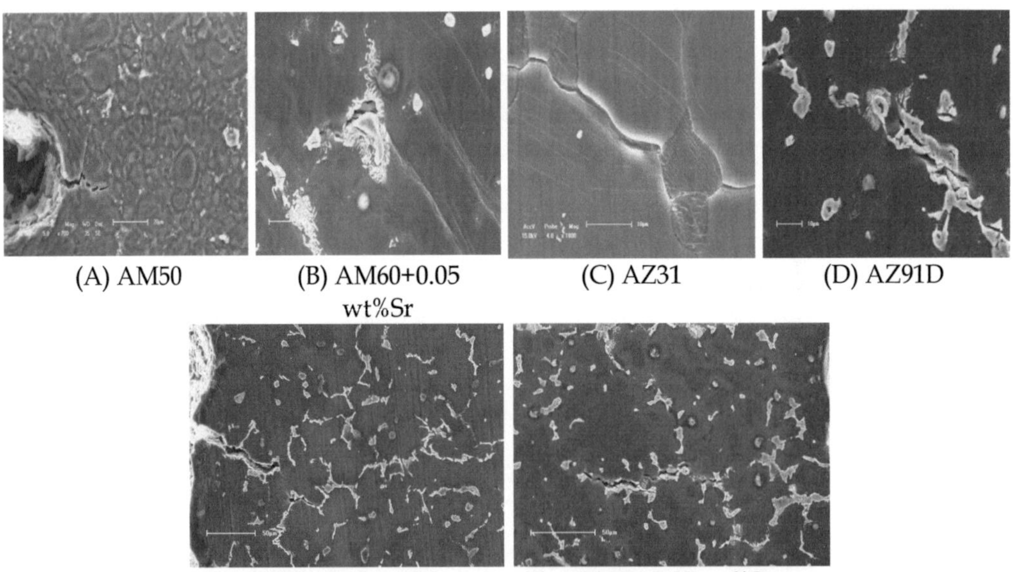

(A) AM50 (B) AM60+0.05 wt%Sr (C) AZ31 (D) AZ91D

(E) AZ91D+0.10 wt%Ca (F) AZ91D+0.05 wt%Sr

Fig. 7. Micro fatigue cracking images of different cast magnesium alloys at R=0.1

crack propagation path in front of a α-Mg grain resulted a deflective angle then around the boundary of this grain after N=5060 cycles as shown in Figs. 8C, this fatigue crack propagation direction evolved gradually a macroscopical crack after N=5222 cycles as shown in Figs. 8D. As the fatigue crack propagation test of cast AM50 alloy, the fatigue crack initiation occurred in the root of notch and propagation was along the boundary of α-Mg grain or interface between α phase and β phase as shown in Figs. 7C and 7D. In addition, the fatigue crack initiation and propagation accompanied always by the obvious plastic deformation vestiges, where occurred mainly on α-Mg phase surface as shown in Figs. 7B and 7C.

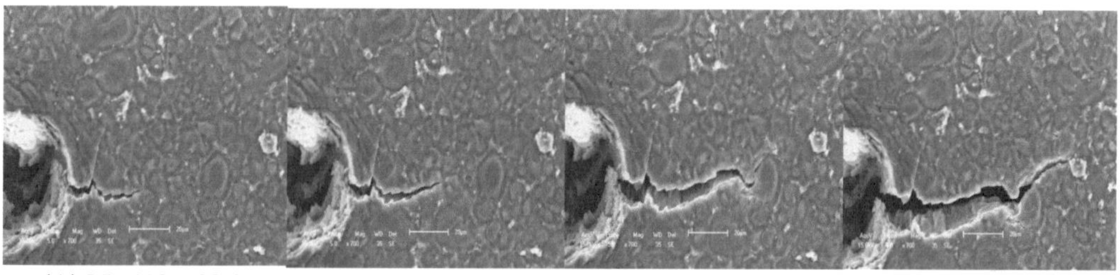

(A) N=4181, 33.9 μm (B) N=4612, 44.9 μm (C) N=5060, 70.5 μm (D) N=5222, 95.5 μm

Fig. 8. Fatigue crack evolutive process of cast AM50 alloy at RT.

For majority investigations on the fatigue crack initiation and propagation of cast magnesium alloys, their fatigue cracking mechanism has the similar result at room temperature. For example, the fatigue crack initiation of cast AM60 alloy occurred also at the root of manual notch as shown in Figs. 9A, where accompanied by some plastic deformation vestiges around the fatigue crack. With increasing the cycles, the fatigue crack propagated along the boundary or β-$Mg_{17}Al_{12}$ phase compared with the results in Figs. 9A and 9B. In addition, the incontinuous crack initiated points were gradually linked as a main fatigue crack or to be defined as the Stage II of fatigue crack growth as shown in Figs. 9B with increasing of cyclic numbers. A larger number of studies devolving on microstructural observations have been conducted, however, and these allow us qualitatively to describe fatigue initiation as it is found to take place in the majority metallic materials. Fatigue fractures in these materials originate almost exclusively at internal or external surfaces, the latter being more common. In all materials there are regions of local inhomogeneity, that result in local "softening," or surface defect to cause local stress concentrations. Either or both of these factors can result in localized plastic flow, which, under the action of a cyclical stress (/strain), can produce surface features that bear, in some senses, a resemblance to a crack or flaw. The plastic vestiges around the crack to be about 45° tilted to the applied loading as shown in Figs. 9A, in which indicated that the "extrusions" and "intrusions" occurred on the surfaces of α-Mg grains. We note that this nucleation stage, Stage I, of fatigue fracture is crystallographic in nature; that is, it is dictated by flow, rather than by tensile fracture, considerations. As a consequence, the initial fatigue crack plane normal is not parallel to the principal tensile axis, and the nucleated crack propagates initially at an angle other than 90° to this axis. Following these nucleation and initial crack-propagation events, slow crack growth (Stage II of fatigue) ensues after the Stage I crack has grown to some critical size determined by material mechanical properties and the applied stress level and state as shown Figs. 9B.

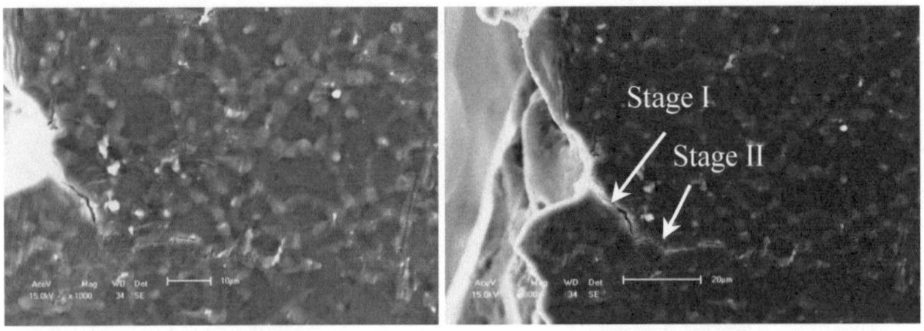


(A) N=53722 (B) N=62102

Fig. 9. Fatigue cracking images of cast AM60 alloy under applied stress of 120 MPa at R=0.1.

The relations of all fatigue crack growth lengths versus the cyclic numbers can be expressed by $d\ell / dN \propto \ell$ when the applied stress is a constant. As the typical curves of cast AM50, AM60B and AZ91 alloys, the fatigue crack growth length shown in Figs. 10 at the different applied stress levels in a single log-coordinate system. For the cast AM50 alloy, the micro cracks growth lengths of $\log\ell$ are almost proportional relations to the cycles under the different applied stresses and the crack growth lengths can be effectively measured from 2 μm to 400 μm by using the SEM *in-situ* observation technology as shown in Figs. 8. With increasing of the maximum applied stress in these crack growth curves, the slope of fatigue crack growth curves is also to become larger and larger as shown in Figs. 10A. If we let to do the first derivative to the crack growth length in these curves, the relation of $d\ell / dN \propto \ell$ (the maximum applied stress is a constant) can be obtained. In addition, there is a slight difference in the slope of fatigue crack growth curves for both AM60B and AZ91 alloys under the same applied stress of 120 MPa, which indicated that the fatigue crack growth rate is not only dependence on the applied stress levels, but also dependence on the mechanical properties of cast magnesium alloys as shown in Figs. 10B.

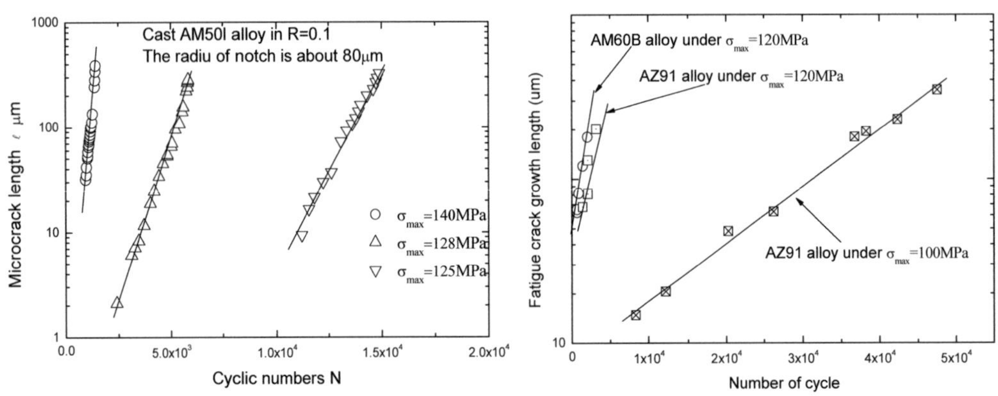


(A) Crack length versus cycles of AM50 alloy

(B) Crack length versus cycles of AM60B and AZ91 alloys

Fig. 10. Crack growth length versus cyclic number of cast magnesium alloys.

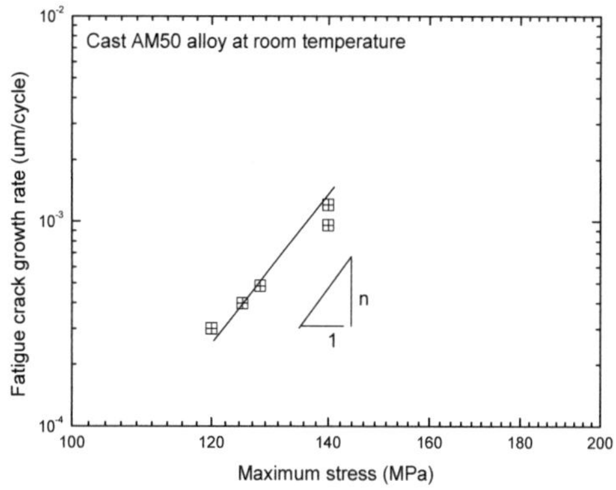


Fig. 11. $d\ell / dN$ versus $\sigma_{\max}$ stress

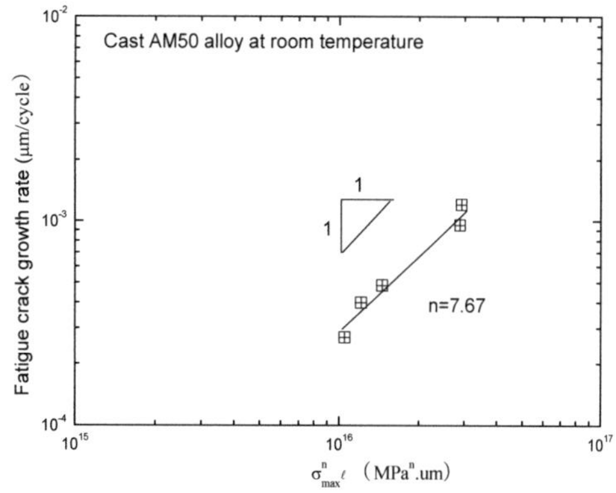


Fig. 12. $d\ell / dN$ versus $\sigma_{\max}^{n}\ell$

As the curves of the fatigue crack growth at different conditions, if the crack growth length is as a constant, such as $\ell = 1$, we can obtain the linear relation of crack growth rate versus the maximum applied stress in the double logarithmic coordinates as shown in Fig. 11. The slope of linear curve is about 7.67 that it indicates that the fatigue crack growth rate can be quantitatively and effectively expressed by the term of $\sigma_{\max}^{7.67}$ when the fatigue crack growth length is one. Therefore, synthetically considered the effect of fatigue crack growth length and applied stress level on the fatigue crack growth rate, the curve about fatigue crack propagation tests of cast AM50 alloy at RT can be plotted as shown in Fig. 12. The slope of the curve is about 1.050, which means that the fatigue crack growth rate of cast AM50 alloy can be quantitatively and effectively characterized by the term of $\sigma_{\max}^{n}\ell$, where $n$=7.67. Therefore, the relationship between the fatigue crack growth rate and the term is written as following:

$$d\ell / dN = C\sigma_{\max}^{n}\ell \tag{4-1}$$

where $C$ is $3.16\times10^{-20}$ (cycle/(MPa)$^n$) for cast AM50 alloy at room temperature. The constants ($C$, $n$) in Eq. (4-1) depend strongly on the experimental conditions and the mechanical properties of cast magnesium alloys, which are the cast magnesium alloys, temperatures, environments and applied loading types, such as push-pull, bending loading, and the stress ratio $R$ etc., respectively. Especially the slight change of the index $n$ will cause the larger change of the constant $C$. Therefore, the higher precision of experimental data is possible for SEM *in-situ* measurements of the fatigue crack growth tests of cast magnesium alloys.

Based on the both parameters of applied stress level and measurable crack growth length in this empiric equation, which involves to the ultimate capable measurement of fatigue crack length $\ell_0$ to allowing the fatigue crack growth length $\ell_n$ of structural element, we can simply predicate the fatigue crack growth life of cast magnesium alloys as following:

$$N_c = \int_{\ell_0}^{\ell_n} \frac{d\ell}{C\sigma_{\max}^{n}\ell} = \left(\frac{\log \ell}{C\sigma_{\max}^{n}}\right)\Bigg|_{\ell_0}^{\ell_n} = \frac{\log \ell_n - \log \ell_0}{C\sigma_{\max}^{n}} \tag{4-2}$$

Above mentioned the estimation of fatigue crack growth life, there is an obvious benefit that the fatigue crack growth law is rather simple because it is based on the engineering stress amplitude or maximum stress at the certain stress ratio and the measurable crack growth length. In addition, as a general engineering fatigue life of materials, it is able to include to two parts of fatigue life, one is the fatigue crack initiated life and another is the fatigue crack propagation life. The farmer is mainly to be decided by the fatigue experiments of smooth specimens or to be deduced by *S-N* curves of smooth specimens. The latter is mainly to be estimated by the fatigue crack propagation experiments with the notch specimens. And it is necessary that considering the effects of experimental and environmental conditions including to the effect of modeling method on the fatigue life of cast magnesium alloys.

### 2.3.2 Fatigue cracking mechanism at elevated temperature

As above mentioned results that the fatigue crack initiation and propagation behavior of cast AM50, AM60, AZ91 alloys are at room temperature. In this section, we introduce simply the effect of the elevated temperature on the fatigue crack initial and propagated mechanism of typical cast magnesium alloys. For example, Figs. 13 shown that the typical fatigue cracking features of cast AM50 alloy at the different elevated temperatures but under the same applied stress of 125 MPa. In the Figs. 13A, the fatigue crack initiation occurred still at the root of notch but the early stage of crack propagation is along either the boundary of α-Mg grain or to cleave the α-Mg grain in front to the crack tip. The fatigue cracking mechanism in the microscopically zone is analogous to the quasi-brittle or quasi-ductile (intervenient brittle and ductile) fracture mechanism of engineering materials. At room temperature, we did not find that the α-Mg grain of cast AM50 alloy was cleaved. However, the fatigue crack propagation is either along the boundary of α-Mg grain or to cleave the α-Mg grain as shown in Figs. 13C. In addition, the fatigue crack propagation mechanism of cast AM50 alloy at the elevated temperature indicated that the fatigue branch crack was found as shown in Figs. 13B. This means that the fracture mechanism of cast magnesium alloy at the elevated temperature which is dominated by the couple of Mode I and Mode II, which differs obviously from that which is main fracture Mode I at room temperature in microscopically zone. Although the cyclic force acting normal to the crack surface serves to open up the crack and to propagate it in a direction normal to the tensile stress, a shear component of stress is applied normally to the leading edge of the crack, which propagates in a direction parallel to the sense of the applied stress to be caused by the plastic deformation in local region at the elevated temperature state. Therefore, the branched probability of fatigue crack is along about 45° tilted to the crack propagation direction as shown in Figs. 13B. This is because there is the maximum shear component of applied normal stress in this direction. Corresponding effect of the elevated temperatures on the fatigue cracking mechanism of cast magnesium alloys has analogous to composite fractures Mode I/II. Therefore, the effect of the elevated temperature on the fatigue cracking mechanism of cast magnesium alloys can be not ignored. This effect of the elevated temperature on the fatigue branched cracking can be contributed by the hardness of β-$Mg_{17}Al_{12}$ to have a softness trend so that the plastic deformation mismatch in the boundary or interface becomes decreasing or weak. This is a competitive result of the interface strength and the fracture strength of α-Mg grain.

(A) 100°C , N=1336 (B) 150°C , N=5090 (C) 180°C , N=5372

Fig. 13. Fatigue cracking characterizations of cast AM50 alloy at the elevated temperatures under the maximum applied stress of 125MPa at R=0.1.

All typically measured results of fatigue crack propagation length of cast AM50 alloy at the different elevated temperatures according to the projection to the loading direction can be plotted in Figs. 14. These curves of fatigue crack growth length at the different elevated temperatures are still similar to the linear trends under the different applied stress levels. It means that the fatigue crack growth rate ( $d\ell / dN$ ) is also the direct proportion the fatigue crack growth length ( $\ell$ ) when the applied stress is a constant. Therefore, if the fatigue crack growth length is a constant, the relationship between the fatigue crack growth rate and the applied stress amplitude at the different elevated temperatures can be plotted as shown in Fig. 15. These slopes of curves of cast AM50 alloy at the different elevated temperatures slightly decreased with the increasing the elevated temperatures, which are 7.67 at room temperature, 6.63 at 100 °C, 6.55 at 150 °C and 6.20 at 180 °C. It means that cast magnesium alloy was intenerated with increasing the temperatures. Synthetically considering the interactional effect of the fatigue crack growth length and the applied stress levels at the different temperatures, the fatigue crack growth rate can be also characterized by the term of $\sigma_{\max}^{n}\ell$ , which is similar to that at the room temperature, as shown in Fig. 16. It is clearly seen that the elevated temperatures caused not only the slight change of index of fatigue crack growth rate but also main influence the constant of C in Eq. (4-1). The higher of the elevated temperature is, the larger of C is as shown in Fig.18. That is, the fatigue crack growth rate at the higher elevated temperature is faster than that at the lower elevated temperature because of the slight difference in the fatigue cracking mechanisms.

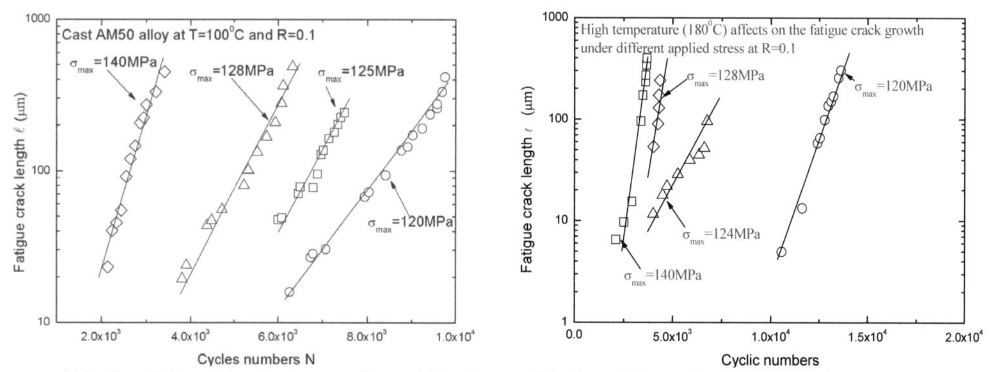


(A) Crack length versus cycle at 100°C (B) Crack length versus cycle at 180°C

Figs. 14. Crack growth length versus cyclic number at the elevated temperatures

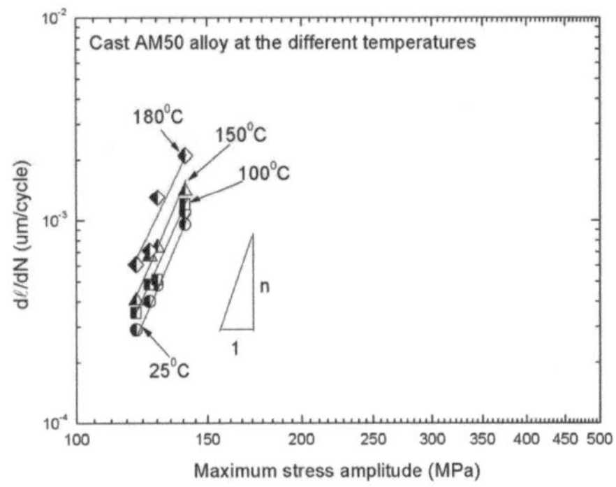


Fig. 15. $d\ell / dN$ versus $\sigma_{\max}$

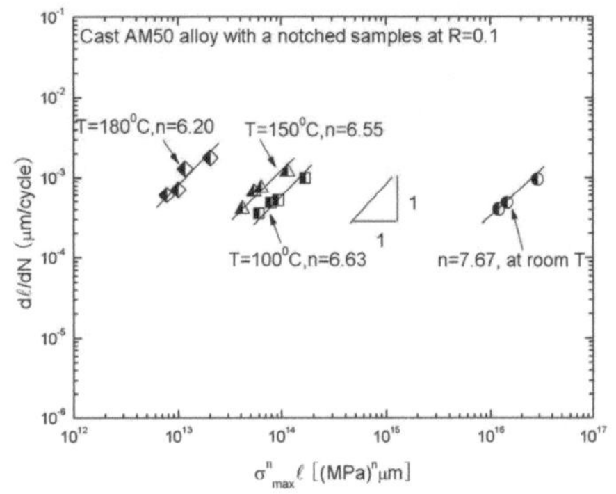


Fig. 16. $d\ell / dN$ versus $\sigma_{\max}^{n}\ell$

At the different elevated temperatures, the dominated index $n$ in fatigue crack growth rate can be estimated by the linear function of $n = 7.78 - 9 \times 10^{-3} T$ as shown in Fig. 17. The variations of power index $n$ versus different elevated temperatures can be explained by the difference of yielding stress and mechanism of small fatigue crack growth at different temperatures. And the constant in Eq.(4-1) versus the elevated temperatures can be estimated by the $\log C = -24.56 + 3.62 \log T$ as shown in Fig. 18. Therefore, the fatigue crack growth rate of cast AM50 alloy at the different temperatures can be also predicated by Eq.(4-1).

Measurements or estimation of the crack growth rates are useful for an engineering design, and they also add to our understanding of the fatigue process. For example, knowledge of the Stage II as shown in Figs. 9B crack-growth arte and the cast magnesium alloys' fracture toughness permits an estimation of the number of Stage II cycles prior to catastrophic final fracture. Thus, for a cast magnesium alloy subject to LCF ($N \leq 10^3 \sim 10^4$ cycles), for which Stage II occupies a majority portion of the cast magnesium alloy's life, the number of fatigue cycles it can withstand prior to failure can be approximated. Moreover, as discussed in the above section of microstructural features of cast magnesium alloys, many structural members contain preexisting surface flaws or cracks that can be precursors to fatigue (tensile) failure, and which eliminate the necessity of nucleating a fatigue crack. For these, knowledge of the critical flaw size and its geometry also allows estimation of fatigue lifetime.

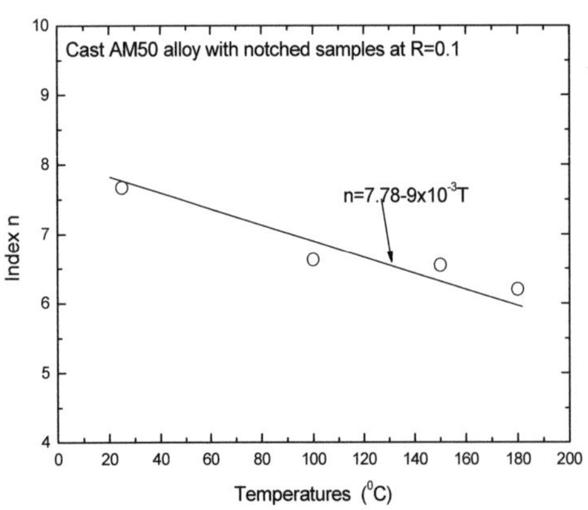


Fig. 17. Index $n$ in Eq. (1) versus $Ts$

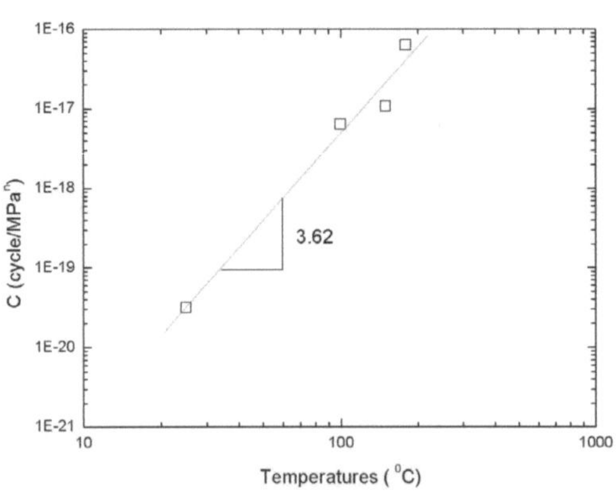


Fig. 18. Constant $C$ versus $Ts$

Measurement of Stage II crack growth rates are commonly performed in many laboratories. A pre-notch sample of the kind used for fracture-mechanics tests (cf. in next section of 2.4) suffices also for the measurements of fatigue-crack growth rate. The sample is typically subjected to a fixed stress (or in some cases, strain) amplitude at a specified mean stress or stress ratio ($R$), and the crack growth length is monitored as a function of the number of cycles. Crack growth length can be measured in a variety of ways, including direct measurement with an optical microscope or by measuring the electrical resistance across the fractured portion of the sample or indirect measurement with the plastic replication method, accompanied by a suitable calibration procedure. These results obtained from such testing are illustrated in Figs. 8, Figs. 10 and Figs. 14. Except to recognize the fatigue crack growth rates of materials can be expressed by the above term ($\sigma_{\max}^{n}\ell, \Delta\sigma^{n}\ell, \sigma_{m}^{n}\ell$, $\Delta\sigma$ is a stress range and $\sigma_m$ is a mean stress etc.), the fatigue crack (Stage II) growth is driven principally by the same kinds of forces that are responsible for tensile fracture. The driving force of crack growth is dependence on the particular cyclical stress history and stress intensity, which scales with the product of stress and the square-root of crack length. When extended to fatigue fracture, the same approach is taken, with the exception that, in recognition of the necessity of a cyclical stress for fatigue, the stress range $\Delta\sigma$ substitutes for $\sigma$ in the stress-intensity factor (SIF). In fact, a larger number of studies have shown that, for a given material and stress ratio, Stage II crack growth rates are a unique function of $\Delta K(\sim \Delta\sigma(\ell)^{1/2})$, and that over an appreciable range of this variable, $d\ell / dN$ is related to it by Paris [40]

$$d\ell / dN = A\Delta K^{m} \tag{4-3}$$

Where $A$ is a constant that depends on the material and the stress ratio, and $m$ is an empirical constant (2≤$m$≤4, usually) deduced from crack-growth rate measurements.

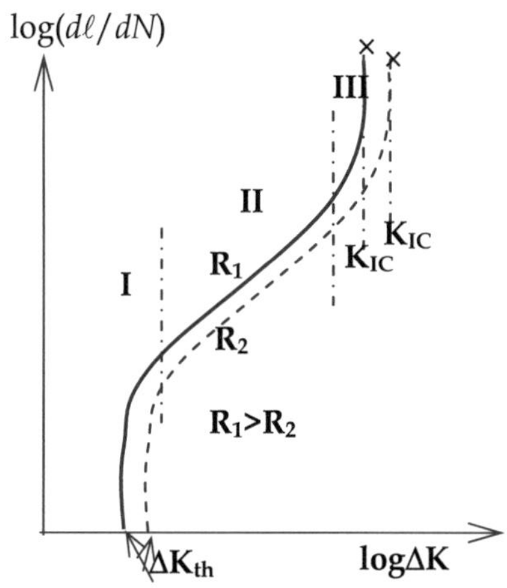


Fig. 19. Schematic of crack-growth rate as a function of the cyclical stress intensity factor for different $R$ values ($R = \sigma_{\min} / \sigma_{\max}$)

Although $d\ell / dN$ of materials may be uniquely as a function of $\Delta K$, it must be emphasized that Eq. (4-3) applies only over a portion of the $d\ell / dN - \Delta K$ curves as similar to above mentioned method which used by the term of $\sigma_{max}^{n}\ell$. This is made clearer by consideration of Fig. 19, which shows both schematic plots of $d\ell / dN - \Delta K$. From these figures we see that at low $\Delta K$ values, crack growth rates are very low (Stage I, as shown in Figs. 9). Indeed, it appears there is a threshold value of $\Delta K$ $(= \Delta K_{th})$ below which Stage II fatigue cracks cannot realistically propagate. This value, $\Delta K_{th}$, represents inherently safe design against fatigue fracture. Unfortunately, $\Delta K_{th}$ is so low in comparison to critical stress intensities (typically 5~10 per cent of them) that use of engineering materials at such low $\Delta K$ values represents a great restriction on their effective utilization. Current engineering design, therefore, is to accept the presence of fatigue cracks in most structures, to assume that some of them will propagate, but to insure, also, that they will not grow to catastrophic length during the intended lifetime of the part. In contrast, if the critical crack growth rate is defined as $10^{-10}$ m/cycle (ASTM: E647, 1998) [41], we can decide the value of $\Delta K_{th}$ or the term of $\sigma_{max}^{n}\ell$.

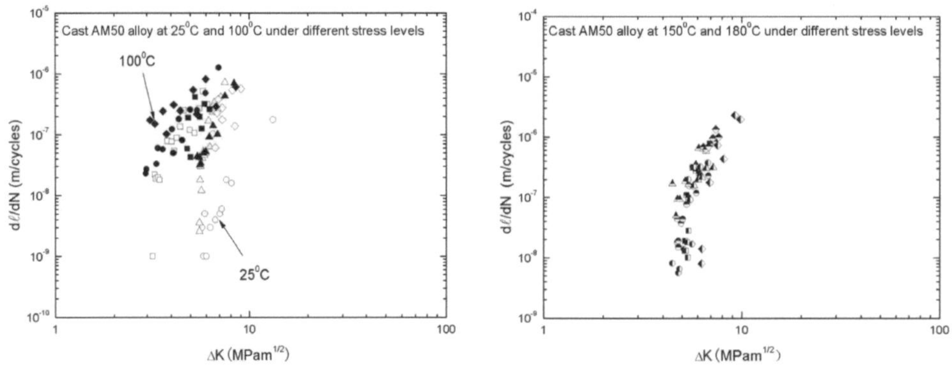


(A) $d\ell / dN - \Delta K$ at 25°C and 100°C (B) $d\ell / dN - \Delta K$ at 150°C and 180°C

Fig. 20. Crack growth rate as a function of SIF at the different temperatures.

Therefore, based on the results of fatigue crack propagation tests at the different elevated temperatures and SIF, the experimental results were plotted as shown in Figs. 20. These results have a larger scatter at RT and 100 °C than that at over than at 150 °C when the maximum applied stress amplitudes are 120 MPa, 125 MPa, 128 MPa and 140 MPa, respectively. It means that the fatigue crack growth rate at both RT and 100 °C can not be characterized uniquely by SIF as shown in Figs. 20A although the fatigue crack growth rate can be uniquely characterized by the term of $\sigma_{max}^{n}\ell$ at the different temperatures as shown in Fig. 16, respectively. However, these results indicated that the fatigue crack growth rate can be simply estimated by SIF at the elevated temperatures from 150° to 180°C as shown in Figs. 20B, which is because the fatigue crack growth mechanism is similar at the higher temperatures. In addition, it is still difficult to evaluate the fatigue crack growth rates of cast AM50 alloy uniquely and reliably based on the SIF at different elevated temperatures. With increasing of the maximum applied stress amplitude at the same temperature or with increasing of the temperature at the same applied stress, the fatigue crack growth rates are varied but they do not uniquely depend the $\Delta K$ as shown in Figs. 20A. When the temperature is over than 150 °C, the fatigue crack growth rate at the closer threshold $\Delta K_{th}$, which is defined as $10^{-10}$ m/cycle, can be uniquely decided by SIF with some scatters about it as shown in Figs. 20B.

Stress parameters needed to characterize the push-pull fatigue test (and other types of, such as rotating-beam fatigue test) are related to the maximum ($\sigma_{\max}$) and minimum ($\sigma_{\min}$) stresses the part is subjected to each cycle. These include the mean stress $\sigma_{mean} = (\sigma_{\max} + \sigma_{\min})/2$, the stress range $\Delta\sigma = \sigma_{\max} - \sigma_{\min}$, and the stress amplitude $\sigma_a = (\sigma_{\max} - \sigma_{\min})/2$. Frequently, the stress ratio $R = \sigma_{\min}/\sigma_{\max}$ is also used parametrically; clearly, $R$ is redundant provided $\sigma_{mean}$ and $\sigma_a$ are known. For the rotating-beam fatigue test, we see that $\sigma_{mean} = 0, \Delta\sigma = 2\sigma_{\max}, \sigma_a = \sigma_{\max}$ and $R$=-1.
Only a few of structural parts prone to fatigue failure experience cycles that are simulated by alternating compression and tension stresses of equal magnitude. Thus, the tests other than the rotating-beam test are better suited for assessing the fatigue resistance of other parts. A cyclical tensile test is often suitable for this purpose. In it, a specified stress amplitude is cyclically imposed on a finite mean stress; a typical stress-time history for such a procedure is shown in Figs. 21.

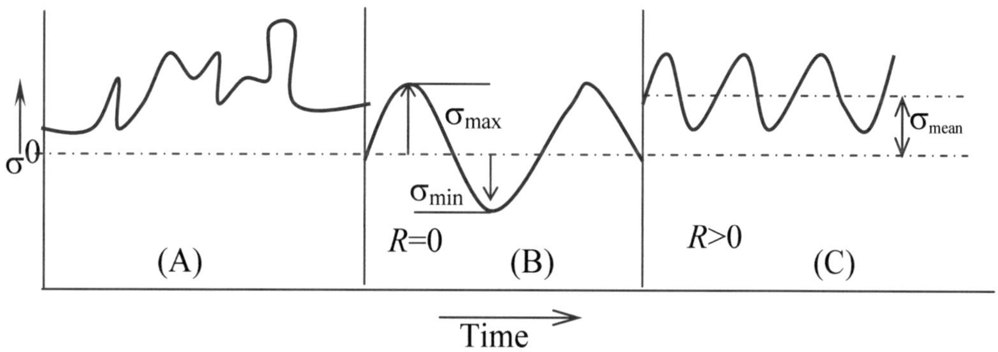


Fig. 21. Characteristic stress-time variations in (A) an engineered structure subject to a positive mean stress on which is superimposed random loading, (B) a rotation beam fatigue test in which the material experiences alternating compressive and tensile stresses of equal magnitude, and (C) a cyclical tensile test in which a time-varying sinusoidal stress is imposed on a constant mean stress.

Because of the important role of plasticity in fatigue damage, it is fundamentally more sounds to assess a cast magnesium alloy's fatigue response under different conditions of a specified cyclically applied strain, rather than stress. Nonetheless, stress-controlled tests are still traditional and are also conveniently performed; the results from them are widely used in engineering design against fatigue damage. Based on the results of stress-controlled tests, it is possible to estimate the relation of the cyclical stress-strain of cast magnesium alloys. For example, during high-cycle fatigue (HCF) (when the number of cycles to failure is very larger (>$10^3$~$10^4$)) the macroscopic stress level is such that the structure as a whole undergoes only elastic deformation, and, in this case, the elastic strain range ($\Delta\varepsilon_e$) is coupled to the stress range by:

$$\Delta\varepsilon_e = \Delta\sigma / E \tag{4-4}$$

where $E$ is the elastic modulus of a cast magnesium alloy. Conversely, in the low-cycle fatigue range, the cast magnesium alloy is typically subject to the plastic strain in macroscopic and microscopic scales. For relatively low values of $N_f$, the plastic strain range is as following:

$$\Delta\varepsilon_p = \Delta\varepsilon_t - \Delta\varepsilon_e \tag{4-5}$$

where $\Delta\varepsilon_t$ is a measurable total strain value by a strain sensor. $\Delta\varepsilon_e$ is calculated by and applied stress range according to the Eq. (4-4). Sometimes, the plastic strain range of some cast magnesium alloys (if the ratio of $\sigma_{\max} / \sigma_{0.2} > 0.75$) is much greater than the elastic one, so that

$$\Delta\varepsilon_t \approx \Delta\varepsilon_p \tag{4-6}$$

Therefore, the plastic strain range is related to applied stress range by knowing the cast magnesium alloys' cyclical hardening response. In the general case, the plastic strain range is still decided by Eqs. (4-4) and (4-5). Thus, the strain is also as a simple function of the term of $\sigma_{\max}^{n}\ell$, such as $\Delta\varepsilon_p^{n'}\ell$. Although elongation of majority of cast magnesium alloys is less than 15%, the accumulated plastic deformation is very important for evaluating the fatigue damage especially in elastic-plastic deformation region. The relationship between the applied stress and response strain of cast magnesium alloy in LCF damage process can be expressed by the Ramberg-Osgood model [42] as

$$\Delta\varepsilon = \frac{\Delta\sigma}{E} + (\frac{\Delta\sigma}{k})^{1/m} \tag{4-7}$$

where $k$ is a constant of the cast magnesium alloy. And $m$ is hardening coefficient of a cast magnesium alloy, in general, $0 < m < 1$ [43].

Cast magnesium alloys behave "differently" when subjected to a cyclical stress-strain environment from the way they do when subjected to a monotonously increasing stress or strain as in a tension test. Knowledge of this cyclic behavior enhances our understanding of the fatigue process. Moreover, data obtained from such studies are useful for engineering design against fatigue fracture. The fatigue behavior of cast magnesium alloys under a cyclical mechanical environment can be investigated by subjecting them to either specified cyclical stress or strain amplitude. The latter is more commonly done and the results from this kind of test constitute the focus of our discussions. If fixed strain amplitude (consisting of some plastic strain component) is imposed on a cast magnesium alloy, the stress range is not fixed but varies with the number of cycles or strain reversals. The cast magnesium alloy may either soften (the stress amplitude decreases with increasing of time/cycles) or harden (when it increases with number of cycles). Moreover, the extent of softening or hardening is a function of the plastic strain range. The behavior of a cyclically hardening material is illustrated in Figs. 22A. Hear we see that the stress amplitude required maintaining the specified strain range increases with the number of cycles, and this is also manifested by an increase in the area of the hysteresis loop (the trace of stress and strain over the course of one cycle). When the magnesium alloy cyclically softens as shown in Figs. 22B, the hysteresis loop becomes smaller concomitantly with the decrease in the stress amplitude necessary to maintain the fixed strain range.

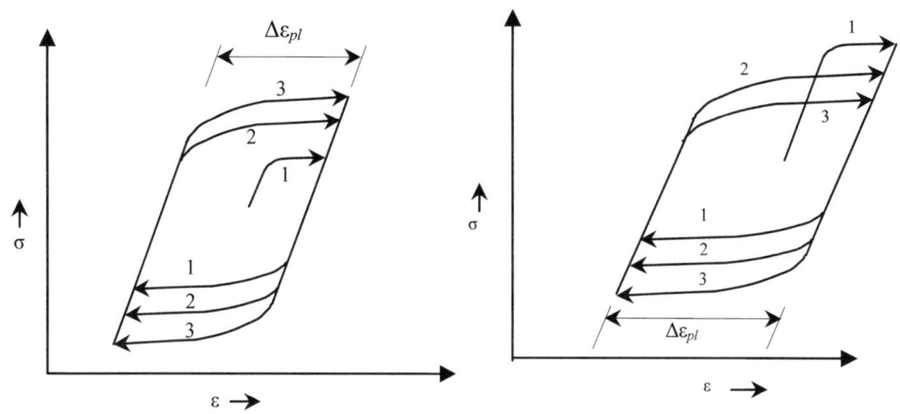


(A) Cyclically hardening curves (B) Cyclically softening curves

Fig. 22. The hysteresis loop of materials

### 2.4 Experimental approach-based on SEM in-situ observation

As an investigation on the fatigue cracking mechanism of cast magnesium alloys relates to the microstructural features, the SEM *in-situ* observation is an impactful approach in the experimental studies. This real detects of fatigue small crack initiation and propagation of cast magnesium alloys were carried out using a specially designed the servo-hydraulic loading testing system in the vacuum chamber of SEM as shown in Figs. 23A. The specimen as similar to the dog-bone shape is put into the jig as shown in Figs. 23B and 23C. Then they are put into the chamber of SEM. To investigate the effect of elevated temperature on the fatigue cracking behavior of cast magnesium alloy, we can use the approach as shown in Figs. 23C and 23D. In this system, the controlled accuracy of temperature is less than ±3°C when the maximum temperature is 800 °C.

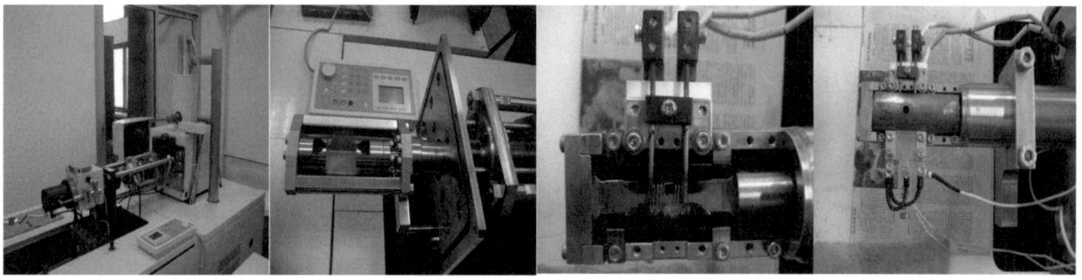

(A) System (B) Jig at room temperature (C-D) Jig at high temperature

Fig. 23. SEM *in-situ* observation and loading with the different jigs systems

All fatigue crack propagation tests are by the loading control at R=0.1 in this chapter. Therefore, the displacements and temperatures have to be performed by two computers. And the experimental data are also recorded in a random time during the fatigue test.

In addition, the system can provide pulsating, such as sine wave etc., loads at different frequencies about general not over than 10 Hz of a certain load capacity. Due there is a deformation on the surface of sample under the applied loadings, it is difficult to observe the clear damaged characteristics in the microscope. Therefore, the special technique was used by Shimadzu Inc. Japan as shown in Figs. 24. Based on the principle of SEM *in-situ* observation, the fatigue micro crack initiation and propagation behaviors of materials become possible in the random cycles. That is, the signal of SEM was directly transferred to a computer via a direct memory access type A/D converter, marking it possible to sample 960×1280×8 bit (or 1920×2560×8 bit) in one frame or over than frames of SEM images successively [13-18,44-48]. You can see why we can obtain the clear images from the presence of deformation specimen surface at less than 0.1 Hz based on the principle and skilled control technology.

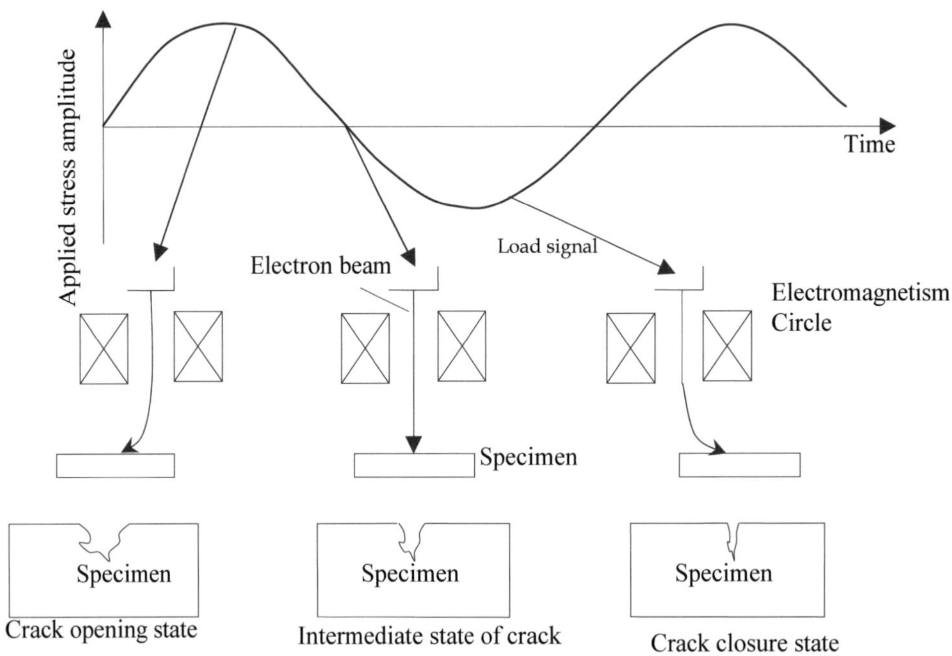


Fig. 24. The principle of SEM *in-situ* observation.

## 3. Effects of spacing and alignment pores on the fatigue cracking behavior of cast magnesium alloys

### 3.1 Fatigue cracking behaviors of cast magnesium alloys with the different spacings and orientations of two pores

As the typical plate with some pores in engineering applications and their especial mechanical properties of cast magnesium alloys, the effects of spacing and alignment pores on the fatigue cracking behaviors have to be considered as an issue in majority studies. Above mentioned the fatigue crack initiation behavior indicated that the effect of stress concentrate at a notch on the fatigue crack initiation can be not ignored. Therefore, the interaction of multi-cracks to occur at the multi-pores on the plate of cast magnesium alloys and the influence of intersectional stress concentrate regions between the multi-pores on the fatigue crack initiation and propagation are also not to avoid. These issues refer the experimental and simulated methods, which can investigate the existent pores whether to contribute faster growth from the origins of crack. In this section, we investigated the effects of different spacing and alignments of manually small pores (as shown in Table 3) on the fatigue crack propagation and fracture behaviors of cast magnesium alloys based on SEM *in-situ* technology and simulation results.

| Materials | Diameters of two pores (mm) | Spacing between two pores (mm) | Alignments of two pores | |
|---|---|---|---|---|
| | | | 90° inclined pores | 45° inclined pores |
| AM60B | 0.50 | 1.00 | √ | √ |
| | 0.50 | 2.00 | √ | √ |
| | 0.50 | 3.00 | √ | √ |
| AZ91 | 0.50 | 1.00 | √ | √ |
| | 0.50 | 1.50 | × | √ |
| | 0.50 | 2.00 | √ | √ |
| | 0.50 | 3.00 | √ | √ |

Table 3. Spacing and alignments of small pores for cast magnesium alloys

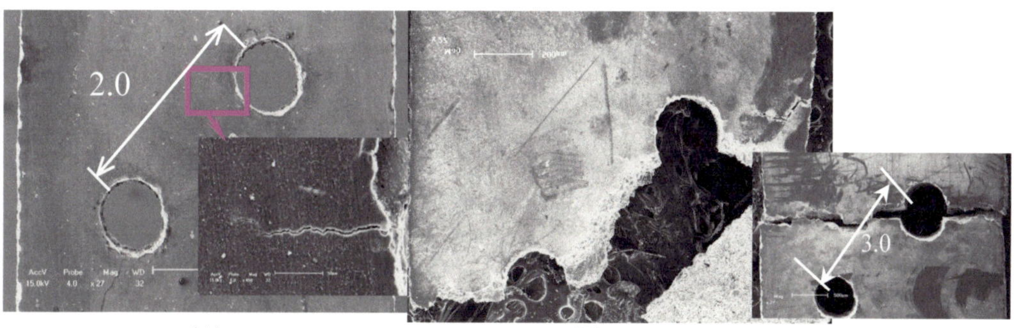


(A) (B) (C)

Fig. 25. Effect of 45° orientations at the two small pores on the fatigue crack growth behavior of cast AM60B alloy. (A) $\sigma_{max}$=100MPa, D=2.0mm, N=5000; (B) $\sigma_{max}$=100MPa, D=2.0mm, $N_f$=9744; (C) $\sigma_{max}$=100MPa, $N_f$=9075, D=3.0mm.

Figs. 25 given the typical fracture cases with two 45° orientations and different spaces under the same stress amplitude and *R*=0.1. The fatigue crack growth path of cast AM60B alloy is obviously different compared with the results both Figs. 25B and Figs. 25C. Before the cycles of about 5000, the early stage of one fatigue crack has occurred at the root of upside pore and the crack growth length is about 150 μm and its propagation direction is about 90° tilted to the applied loading direction. However, the crack propagation direction takes place the deflexion after 5100 cycles subsequently its fracture manner is as shown in Figs. 25B. With increasing the spacing between inclined two pores from 2.0 mm to 3.0 mm as shown in Figs. 25C, the fatigue crack propagation at one pore is different from the result as shown in Figs. 25B. These results indicate that the fatigue crack initiations are still to occur at the stress concentrate regions in closed to the pores and the early stage of fatigue crack propagation is mainly dominated by fracture mode I. These fatigue crack propagation accompanied by the interaction between two inclined-pores. Therefore, the effect of the spacing of two pores on the fatigue fracture is not ignored although the experimental conditions are the approximately same such as the stress amplitude and with the 45° orientations of two pores. For this case, there is a critical spacing to cause the deflexion of fatigue crack propagation direction in the range from 2.0 mm to 3.0 mm, but which is not less than 2.0 mm.

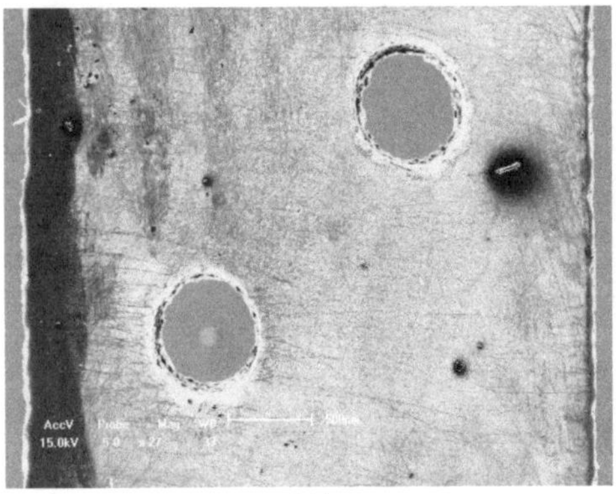


(A)

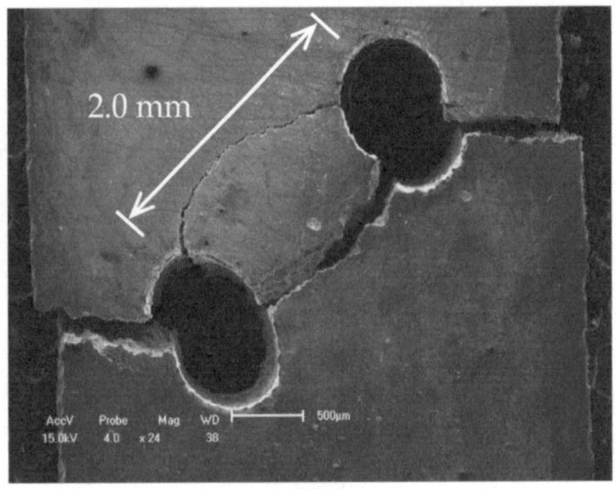


(B)

Fig. 26. Effect of 45° orientations at the two small pores on the fatigue crack growth behavior of cast AZ91 alloy. (A) $\sigma_{max}$=145MPa, D=2.0mm, N=0; (B) $\sigma_{max}$=145MPa, D=2.0 mm, $N_f$=952.

As another typical case with the 45° orientations at the two pores specimen of cast AZ91 alloy, the fatigue crack propagation behavior is shown in Figs. 26A-26B when the applied stress level was changed at R=0.1. The spacing between the 45° orientations at the two small pores is still about 2.0 mm. The result indicates that fatigue crack initiation stochastically occurred at the root of one pore and the fatigue multi-cracks initiation to occur at the roots of two small pores. Thereinto, the fatigue crack in closed edge of specimen propagated along the 90° tilted the applied loading direction and the fatigue cracks between two pores easily produce a coalescence of crack with another one so that the fatigue cracks propagation directions have to deflect an angle in the stress concentration overlapped regions as shown in Figs. 26B. Compared with the results above mentioned cast AM60B, the different mechanical properties of materials (especially the fracture toughness) affect on the difference of fatigue crack propagation path at the approximately experimental conditions as shown in Figs. 25 and Figs. 26. For the specimens with two small pores tilted about 45° orientations to the applied loading axis, the fatigue crack propagation path depends not only on the spacing of two small pores, but also on the fracture toughness of cast magnesium. In contrast, the fatigue crack initiation and propagation behaviors follow as the principle of the maximum strength of material. That is, the fatigue crack initiation occurs at the region of stress concentration and the fatigue crack preferentially propagated along the overlapped region of stress concentration. In addition, the influencing range of two pores with 45° orientations on the fatigue crack growth path is the critical spacing value, which is not over than 2.0 mm when the diameter of pore is about 0.5 mm. If the smaller spacing of two pores is, the greater in the probability of the fatigue crack propagated interaction between two small pores. As the elongation of AZ91 is about 3%, which is less than that of

AM60B, the fatigue crack initiation easily occurs at the many sites around each pore even if there is a strong influence of stress concentrate region. Therefore, these multi-cracks propagated as the manner as shown in Figs. 26B. This reflects the fact that the fatigue crack initiation behavior is a competitive result of the interactions both fracture toughness properties and notch effect of cast magnesium alloys.

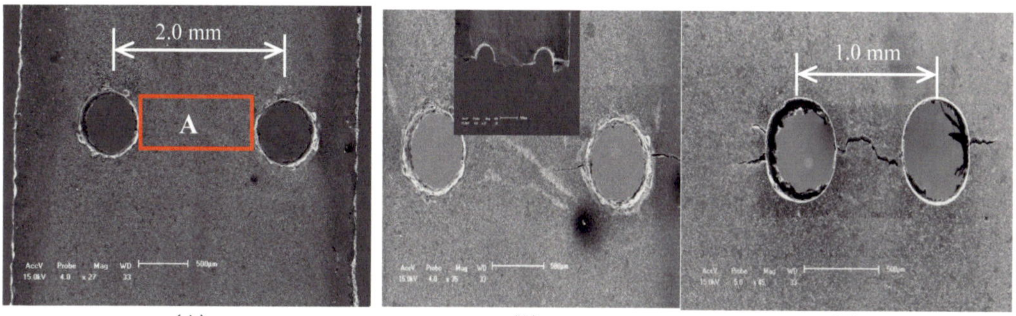


(A) (B) (C)

Fig. 27. Effect of 90° orientations at the two pores on the fatigue crack growth behavior of cast AZ91 alloy at R=0.1. (A) $\sigma_{max}$=100MPa, N=0; (B) $\sigma_{max}$=100MPa, N=52464; (C) $\sigma_{max}$=100MPa, N=38543.

Figs. 27A to Figs. 27C show the fatigue crack initiation and propagation behaviors of cast AZ91 alloy for the 90° tilted to the two small pores and the spacing of about 2.0 mm, 1.0 mm, respectively. These results indicated that the fatigue crack initiation occurred still at the root of notch in each pore. Due the spacing between two pores is greater than the distance to edge of sample, the fatigue crack propagated preferentially to the edge of sample as shown in Figs. 27B. However, if the spacing between two pores is less than 2.0 mm as shown in Figs. 27C, the fatigue crack propagation direction is preferentially to the insider between the two small pores (to refer the mark A zone in Figs. 27A). And the fatigue cracks to occur at the sides between the two pores were easily linked a main fatigue crack so that it cause rapidly the fatigue fracture of specimen as shown in Figs. 27C. That is, the fatigue crack initiation and propagation behaviors depend strongly on the spacing between two small pores and on the two pores with 90° or 45° tilted to the applied loading direction.

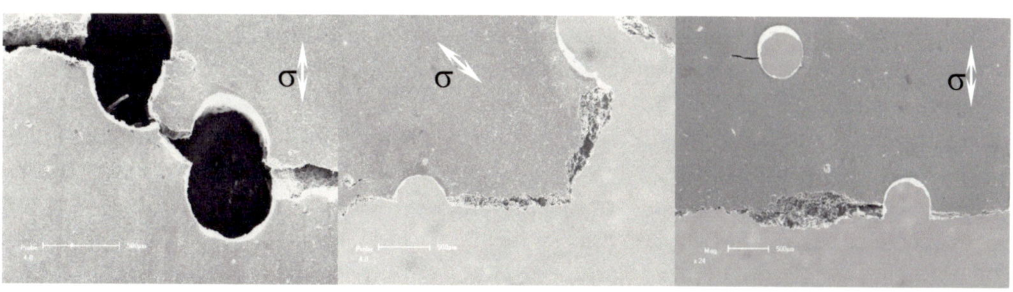


(A) D=1.0 mm, $N_f$=2154 (B) D=2.0 mm, $N_f$=4120 (C) D=3.0 mm, $N_f$=29932

Fig. 28. Cast AM60B alloy with the different spacing of 45° orientations at the two pores under $\sigma_{max}$=120MPa, $R$=0.1.

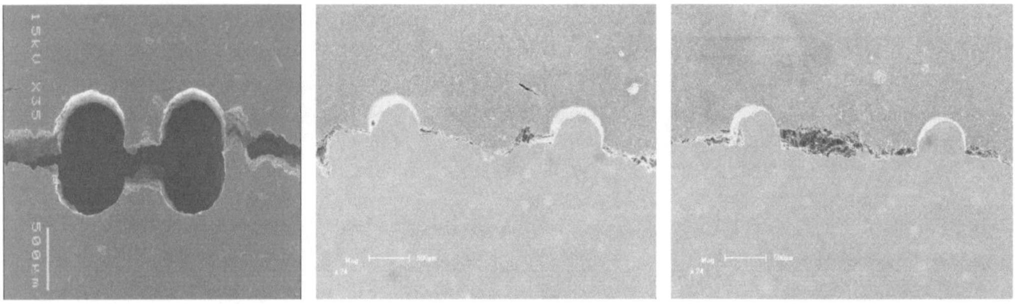

(A) D=1.0 mm $N_f$=11446 (B) D=2.0 mm $N_f$=10005 (C) D=3.0 mm $N_f$=25202

Fig. 29. Cast AM60B alloy with the different spacing of 90° orientations at the two pores under $\sigma_{max}$=120MPa.

To validate the critical spacing value of two pores above mentioned 2.0 mm, Figs. 28 and Figs. 29 show the experimental evidences which are the effective results of different spacing and alignments of cast AM60B alloy under the same applied stress amplitude, respectively. For the specimens with the 45° orientations at the two pores as shown in Figs. 28, the effect of spacing between two pores on the fatigue fracture is obvious. In the Figs. 28C, the final fracture of specimen did not cause the coalescence of cracks at one pore with another pore when the spacing between two pores is over than 3.0 mm. But the spacing between two pores is less than 2.0 mm, the fatigue crack occurred at the two pores finally linked with each other. Therefore, we can obtain that the fatigue life increases with the increasing of spacing with the 45° orientation at the two pores. This reflects the fact that the interaction of fatigue cracks affects on the fatigue life. Likewise, when the alignment of two pores was changed as the 90° orientation as shown in Figs. 29, the experimental results indicated that the fatigue life of cast AM60B alloy is also affected by the spacing between two pores. And the critical value of spacing between two pores is approximate 2.0 mm. That is, when the spacing is over than 3.0 mm, the effect of alignment with two pores on the fatigue life of cast AM60B can ignore although there is slight difference in their fracture lives. One of reasons about this difference focus on the fatigue crack initiation life because of the quality of drilled pores. For example, the effect of the quality of drilled pores on the fatigue life of cast AZ91 alloy shown in Figs. 30. Although there are the same experimental conditions such as the spacing and alignment of two pores as well as under the applied stress level, the fatigue life exists in the slight difference after $5.25\times10^4$ cycles. This is because the fatigue crack initiation occurs preferentially at the rather bigger pore as shown in Figs. 30A. And if the fatigue crack occurs synchronously at the edges of two pores, the fatigue cracks cause easily a coalescence of cracks with another one, especially like as the rapidly linked cracks in the spacing between two pores as shown in Figs. 30B.

Alternatively, for 45° orientations at the two pores with the spacing (which is 1.5 mm, 2.0 mm, 3.0 mm, respectively) of cast AZ91 alloy, the fatigue life and fracture manners under the different applied stress amplitudes are respectively shown in Figs. 31. The fatigue cracks propagation path difference of cast AZ91 alloy with two pores specimens is very obvious. The interaction of fatigue cracks did not occur when the spacing is 3.0 mm.

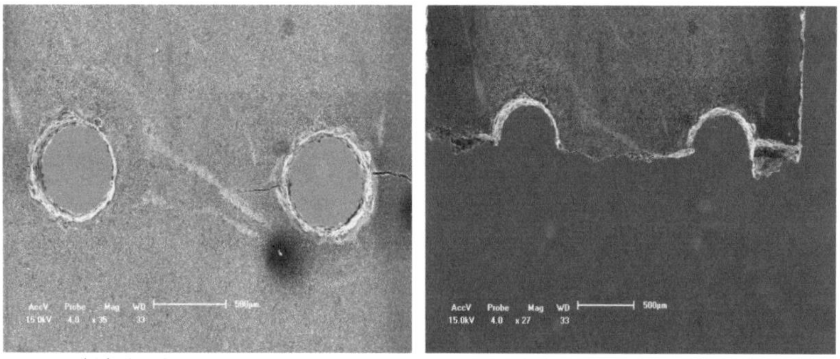

(A) D=2 mm N=52464 (B) D=2 mm $N_f$=52522

Fig. 30. Cast AZ91 alloy with the same spacing of 90° orientations at the two pores under $\sigma_{max}$=100 MPa.

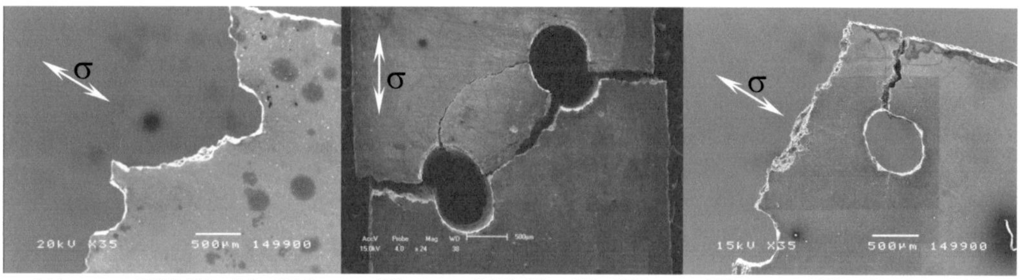


(A) $\sigma_{max}$=170MPa, $N_f$=236 (B) $\sigma_{max}$=145MPa, $N_f$=952 (C) $\sigma_{max}$=123MPa, $N_f$=3924

Fig. 31. Cast AZ91 alloy with the different spacing of 45° orientations at the two pores (1.5 mm, 2.0 mm, 3.0 mm)

For majority cast magnesium alloys, their fatigue lives are still dependence on the mechanical properties, especially the fracture toughness or elongation (%) of cast magnesium alloys. In general state, the greater of the fracture toughness or elongation (%) of cast magnesium alloys is, the longer their fatigue lives are as shown in Fig. 32. For example, the elongation of cast AM60B alloy is about 10% and the elongation of cast AZ91 alloy is about 3% so that the fatigue life of the former is longer than that of the latter under the same experimental conditions. This is because the brittle property of cast AZ91 alloy causes much easier multi-cracks initiation at the β-$Mg_{17}Al_{12}$ phase or interface both α-Mg grain and β-phase than that of cast AM60B alloy, alternatively the effect of smooth specimen and specimen with a notch on the fatigue life of cast AM50 alloy is also obvious, especially in the higher applied stress levels. However, when the cyclic numbers arrives at the $2\times10^5$, the effect of small notch on the fatigue life is gradually weaken because the notch of cast magnesium alloys depends mainly on the fatigue crack initiation life, which is to occupy a little part of total fatigue life of cast magnesium alloys under the lower stress levels. The investigations results indicated that the fatigue crack propagation life of cast AM50, cast AM60, AM60B and AZ91D alloys occupies approximately about 70%. Therefore, the high cyclic fatigue life of smooth specimen and specimen with a small notch has not almost difference under a low stress level as shown in Figs. 32. These curves indicated still that the relative fatigue life relation among cast magnesium alloys. These accurate relations of fatigue live were validated by cast AM50, AM60, AM60B and AZ91. It is fact that the fatigue life of cast AZ91 alloy with a lower elongation is shorter than that of either cast AM60 alloy or cast AM50 alloy with a higher elongation under the same experimental conditions, especially in higher stress levels. Therefore, the fatigue property of cast AZ91 alloy (3%-elongation ratio) is weak to compare with cast AM60 alloy (7-10% elongation ratio) or cast AM50 alloy (9-11% elongation ratio). This is because the elongation ratio of material relates to the fracture toughness of material. The higher facture toughness of material is, the greater fatigue crack propagation resistance is.

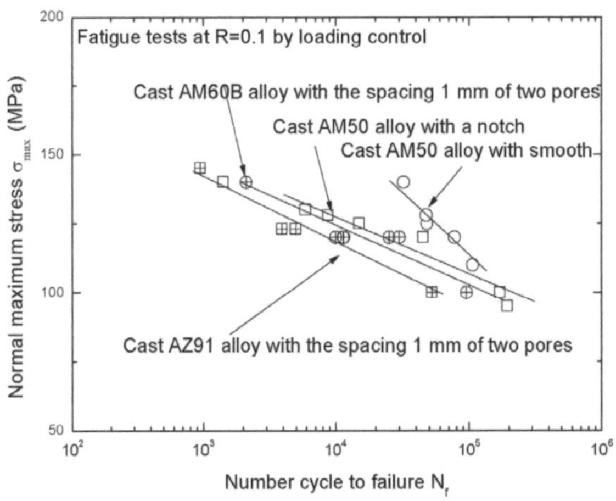


Fig. 32. *S-N* curves of cast magnesium alloys at the different conditions

### 3.2 Validation on the effect of two pores on the cracking behavior by optical displacement microscope

Above mentioned many results about the effects of spacing and orientations with the two pores on the fatigue cracking behavior of cast magnesium alloys, these results indicated that there is a critical spacing value of two pores with the 90° or 45° orientation whether to tack place the coalescence of crack at one pore with another one. To validate the critical value whether is correct and reliable, the tensile tests of cast AM60 alloy with 90° and 45° orientations of two pores were carried out by using the optical displacement microscope, which is usually stretched at a specific rate, and the force required that it is measured to cause an extension of a crack length $\delta\ell$. Force is measured by means of a load cell that is often a calibrated, stiff spring, and the extension is measured often by means of a device called an extensometer. All tensile tests used specimens have the approximate same pore diameter (D) of 0.68 mm and 4D (~2.72 mm) spacing. The tensile speed is about $10^{-3}$ mm/min by the displacement control. These results indicated that there is obvious the stress concentration area evolutive process in prior to the crack initiation at the root of pores as shown in Figs. 33 and Figs. 34. When the applied strain value arrives at the 0.8% in prior to the yield stress as shown in Figs. 33B, the slight stress concentration region (see the white area) occurred at the root of each pore, and the stress concentration area mainly occupied at insider of two pores is greater than that at outsider of two pores as shown in Figs. 33C and 33D. With increasing the applied strain value of $\varepsilon = 3.0\%$, the stress concentration areas at the insider of two pores overlap gradually as shown in Figs. 33E so that the cracks easily accelerate the crack growth in these regions. One clear crack was found in the plastic strain overlapped region when the applied strain value is about 3.7% as shown in Figs. 33F. The total evolutive process of stress/strain concentration on the surface of cast AM60 alloy agrees well with the experimental images based on the SEM *in-situ* observation of fatigue crack propagation tests as shown in Figs. 29.

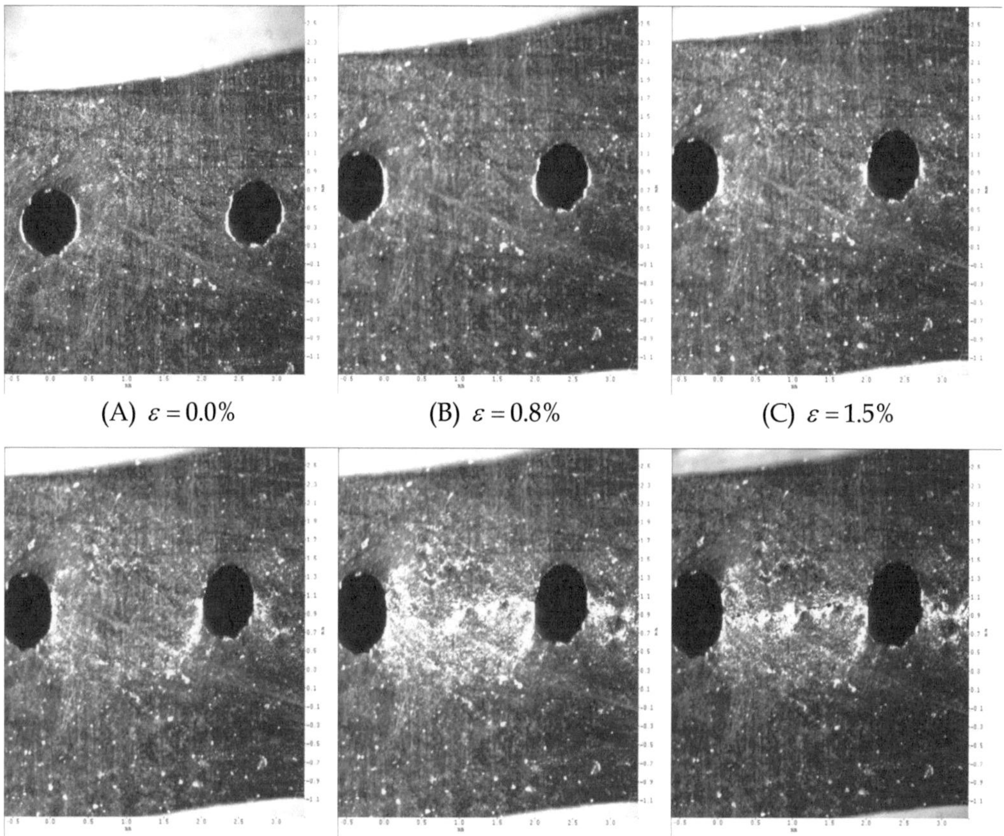

(A) $\varepsilon = 0.0\%$ (B) $\varepsilon = 0.8\%$ (C) $\varepsilon = 1.5\%$

(D) $\varepsilon = 2.2\%$ (E) $\varepsilon = 3.0\%$ (F) $\varepsilon = 3.7\%$

Fig. 33. Evolutive deformation process under the different strains for 90° orientations with the two pores.

As another typical tensile testing images of cast AM60 alloy specimen with two pores at a 45° orientations as shown in Figs. 34, the stress concentration region at the edges of two pores is different from above mentioned results as shown in Figs. 33, especially after the applied strain value being over than 2.0%. This means that the stress overlapped manner will influence on the fatigue crack propagation behavior according to the fracture mechanism of cast magnesium alloys. At the same time, you can see that the stress concentration diffusion process of cast magnesium alloy maybe result the multi-cracks. When the applied strain value increases to about 2.0-3.0% as shown in Figs. 34C and 34D, the stress concentration areas occur at much more overlapping or diffusion part. The crack propagation path has to deflect to the overlapping regions as shown in Figs. 34E and 34F. In addition, the cracks initiation and propagation result the release of stress concentration compared with the change of white regions in both Figs. 34D and 34E. Compared with the results as shown in Figs. 33 and Figs. 34, the effect of two pores orientations on the evolutive process of the stress concentration or on the crack propagation path can be not ignored. The visible tensile and fatigue cracks of cast AM60 alloy occur almost at the root of any pore under the different orientations of two pores. Therefore, the effect of two pores orientations on the crack initiation behavior is rather slight. For example, the crack initiation of cast AM60 specimen with either 90° or 45° orientations occur at the root of any pore when the applied strain is about 3.7% and 4.0%, respectively. However, the effect of orientations of two pores on the strength or fatigue life of cast AM60 alloy can be not ignored. This means that this effect mainly focus on the crack propagation behavior of cast magnesium alloys.

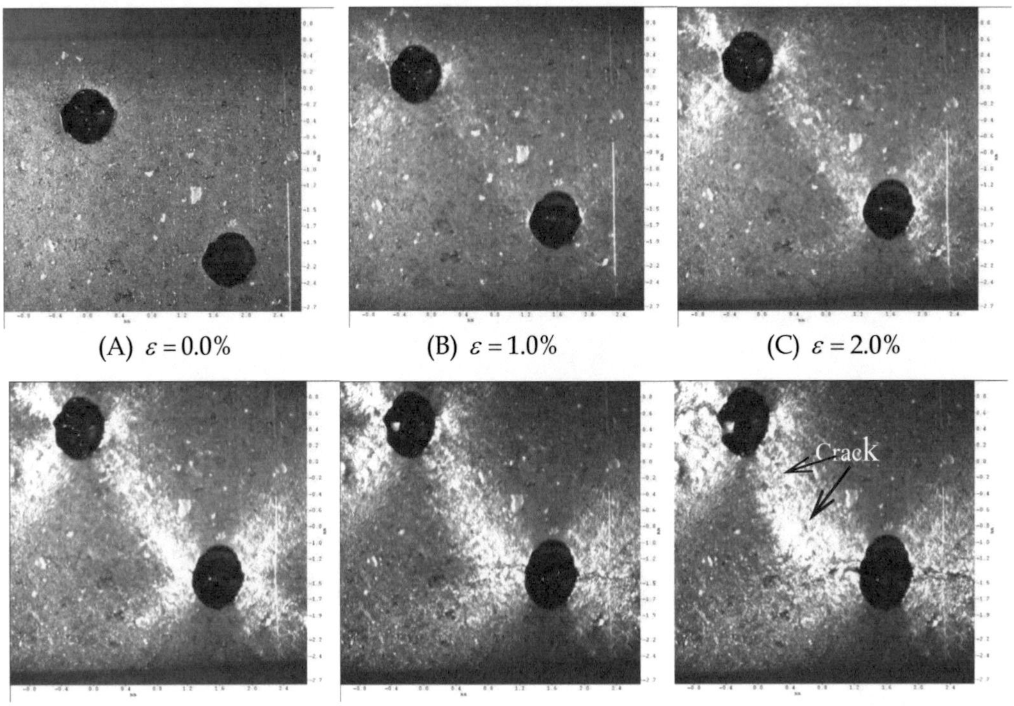


(A) $\varepsilon = 0.0\%$ (B) $\varepsilon = 1.0\%$ (C) $\varepsilon = 2.0\%$

(D) $\varepsilon = 3.0\%$ (E) $\varepsilon = 4.0\%$ (F) $\varepsilon = 5.0\%$

Fig. 34. Evolutive process of plastic deformation under the different strain levels for 45° orientations with the two pores.

To validate the effect of different orientations on the fracture strength of cast AM60 alloy, we give the stress-strain curves of cast AM60 specimens with different orientations of two pores as shown in Figs. 35. These pores can be defined as an initial crack or defect. These results indicated that the effect of different orientations on the fracture strength and fracture toughness of cast AM60 alloy under the static tensile loading is obvious. That is, the fracture strength and toughness of a specimen with 90° orientations are lower than that of a specimen with 45° orientations. This means that the former damages or fractures prior to the latter. As the engineering stress as shown in Fig. 35, the original and strained dimensions are related through $A_0\ell_0 = A_i\ell_i$, where $A_0, \ell_0$ are the original transverse cross-sectional area and sample gage length, respectively, and $A_i, \ell_i$ represent these quantities in the strained condition. As the cross-sectional area decreases with increasing strain, the sample experiences an effective stress greater than that suggested via calculation of stress based on the initial cross-sectional area. We can define an engineering stress as following:

$$\sigma = \frac{F}{A_0} \tag{4-8}$$

When the tensile test of cast magnesium alloy is at low strains, the deference between $A_0$ and $A_i$ can be ignored. Therefore, based on the stress-strain curves in Fig. 35, the effect of cast AM60 alloy with the different orientations of two pores on the tensile properties can be described by the difference between both curves. For example, when the strain value is closed to the 3.0% value, the plastic flow in both curves presents the different stress concentrated characteristics due there is an effect of stress concentrated overlapping region fraction. With increasing the overlapping region fraction, the failure resistance of sample becomes smaller and smaller. On the other hand, the failure possibility of sample with a 90° orientations of two pores is greater than that of sample with a 45° orientations of two pores at the same spacing. The phenomenon is an evidence of a "working softening" effect of material in the stress concentrated overlapping area compared with the images as shown in Figs. 33E and Figs. 34D. At the same time, we can deduce that this is why the fatigue life of a specimen with 90° orientations is lower than that of a specimen with 45° orientations when the spacing of two pores is not over than 2.0 mm above mentioned.

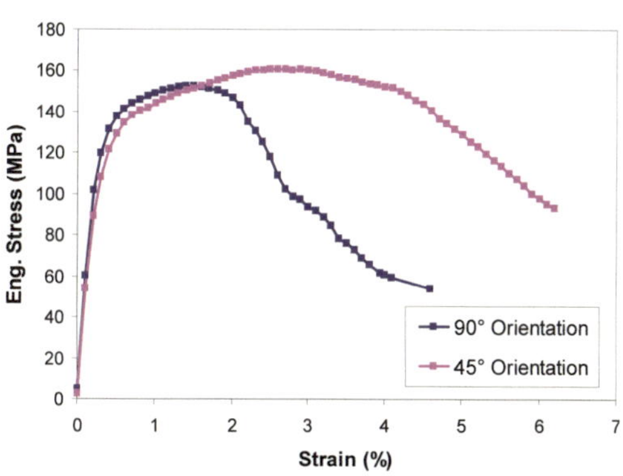

Fig. 35. Stress-strain curves

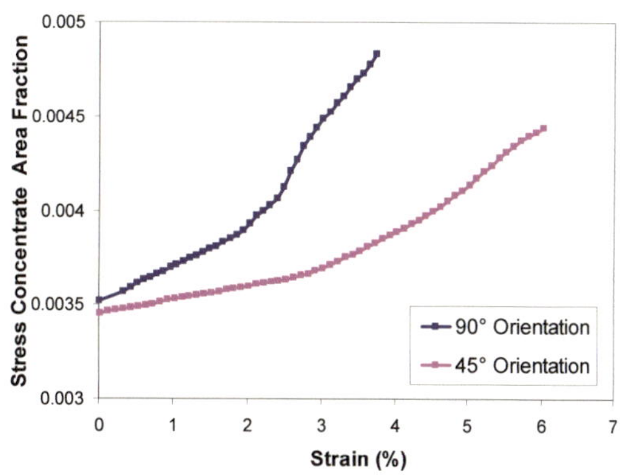

Fig. 36. Stress concentrate diffusion curves

Alternatively, to compare with the difference between the effects of different orientations on the damage of cast magnesium alloys, the relationship curves between the stress concentrated areas near the different orientations of two pores and the applied strain (%) were plotted in Fig. 36. It is clearly seen that the curves of the stress concentrated area fraction versus the applied strain indicates the effect of 90° orientations on the fracture of cast magnesium alloy is much obvious than that of 45° orientations at the same spacing. That is, the stress concentrated overlapping region or diffusion area with 90° orientations of two pores is greater than that with 45° orientations of two pores under the same strain level as shown in Fig. 36. This means that the former easily cause the crack initiation and propagation at these regions because of the theory of fracture mechanics of materials.

Especially it is much obvious for the effect of orientation on the stress concentrated degree when the applied strain value is over than 3%. At the same time, these results validated effectively that the effects of two pores' spacing and orientation on the fatigue crack initiation and propagation behavior are correct based on SEM *in-situ* observations.

### 3.3 Modeling on two pores of cast magnesium alloy

There are several interesting aspects of the critical fracture condition when we consider the effect of two pores with the different spaces and orientations on the failure of cast magnesium alloys. One is that the yield strength and cracking criterion are the same tension as in fatigue of cast magnesium alloys. As mentioned, when the local stress ahead of a crack tip exceeds a critical strength value of cast magnesium alloy, the crack will cause to failure of material. In majority states, the von Mises yield stress can be as the failure criterion value. It is expressed as

$$\sigma_{Miese} = \sqrt{[(\sigma_1 - \sigma_2)^2 + (\sigma_2 - \sigma_3)^2 + (\sigma_3 - \sigma_1)^2]/2} \tag{4-9}$$

The condition states that yielding will not take place for principal stress combination such that the von Mises stress ($\sigma_{Miese}$) near the pores is less than the yield stress of cast magnesium alloy, but that flow will occur if it is greater than the yield stress. Due to the presence of small pore in the plate, the stress states around a pore will cause the stress concentration, stress states transmitted by each other and the cumulative evolvement of plastic strain in the fatigue process of materials. In general condition, we can use the von Mises stress replaces to the uniaxial tensile stress. In fact, the fracture or failure of materials depends strongly on the state of stress. This is why there are the different crack propagation characteristics of the same cast magnesium alloy with different two pores' spaces and orientations in the experimental results.

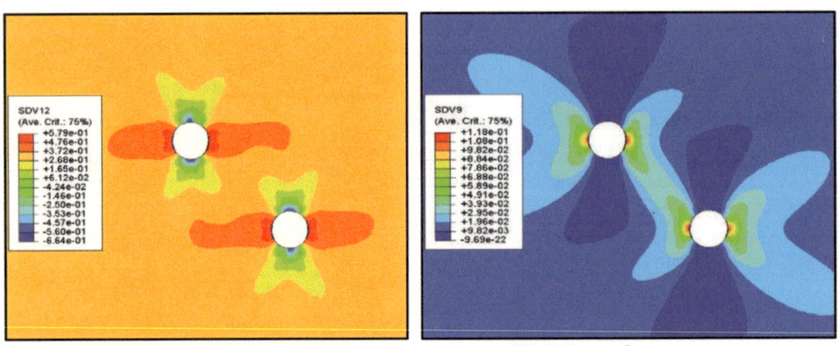


(A) Max von Mises stress (B) Max plastic strain

Fig. 37. von Mises stress and plastic strain distributions around the two pores

To validate the evolutive process of stress concentration region closed to the two pores, we simulated the von Mises stress and maximum plastic strain distributions closed to the two pores. These von Mises stress and plastic strain distributions around the two pores can tell quantitatively us the crack initiation and crack propagation path of material as shown in Figs. 37. The contour plots of stress triaxiality and effective plastic strain at about 2% indicated that the von Mises stress evolutive direction is approximate 90° tilted the applied loading direction, which it causes the main reason of crack initiation as shown in Figs. 37A.

All SEM *in-situ* observations results about fatigue crack propagation tests and tensile tests of cast magnesium alloys with different orientations and spaces of two spherical pores have validated that the crack occurred mainly at the root of pore and the early stage of crack propagation direction was along 90° tilted to the applied loading direction. This means the crack initiation behavior depends strongly on the maximum strength theory of cast magnesium alloys as shown in Figs. 37A. However, with increasing the applied stress level, the strain (plastic strain) value round the two pores is also to increase nonlinearly (to meet with the Ramberg-Osgood model [42]) based on the constitutive equations of cast magnesium alloys as shown in Figs. 37B. Therefore, the crack propagation behavior of cast magnesium alloys with the different orientations and spaces of two pores depends strongly on the overlapping degree of stress concentration or plastic strain, which dominates the macroscopical crack propagation direction of cast magnesium alloys. At the same time, the overlapping degree of stress concentration or plastic strain will cause the fracture mode to transfer from mode I fracture to mode I/II fracture. That is, the shear stress comes from the plastic strain in local area, especially in the overlapping region based on the interaction of stress concentration area around the two pores. For example, when the crack initiation occurs at the root of pore as a sketch map shown in Fig. 38, the stress at crack tip in the mark "•" can be expressed as following:

$$\begin{aligned} \sigma_\theta &= \frac{C_0}{\sqrt{2\pi r}}[K_I(C_1-\cos\theta)\cos\frac{\theta}{2}+K_{II}(3\cos\theta+C_2)\sin\frac{\theta}{2}] \\ C_0 &= \frac{-2\alpha(1-\nu)}{1+3\alpha-4\nu\alpha}, C_1=\frac{-2(1+\nu-2\alpha\nu)}{1+\alpha-2\nu}, C_2=\frac{2(1+2\alpha((1-\nu)}{1+3\alpha-2\nu} \end{aligned} \tag{4-10}$$

Where $K_I, K_{II}$ denote the stress intensity factor of mode I and mode II fractures, respectively. $\alpha$ denotes the modulus ratio of inhomogeneous materials or the inhomogeneous phases. When $\alpha=1$, corresponding to homogeneous material, Eq. (4-10) reduces to the well known result of a mix mode I/II fracture of material:

$$\sigma_\theta = \frac{C_0}{2\sqrt{2\pi r}}[K_I(1+\cos\theta)-3K_{II}\sin\theta]\cos\frac{\theta}{2} \tag{4-11}$$

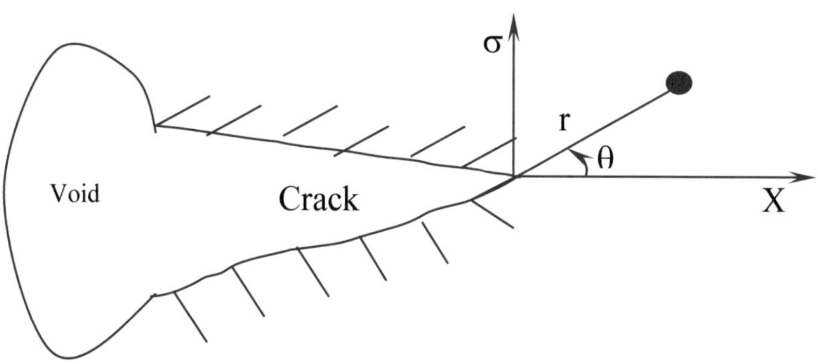


Fig. 38. A mixed fracture mode in the crack tip

From the solution Eq. (4-10) obtained, the angle of maximum tangential stress $\theta_f$ can be determined by using $\partial\sigma_\theta / \partial\theta = 0$ and $\partial\sigma_\theta / \partial\theta < 0$. An 'eligible' interface or boundary, at which the $\sigma_{\theta\max} \ge \sigma_f$ ($\sigma_f$ is the fracture stress of the interface or boundary) is satisfied, will be located at the interface or boundary of cast magnesium alloy. Then the critical stress intensity factor, or the apparent fracture toughness of the cast magnesium alloy, can be established. The crack growth path deflecting main reason should be by local plastic strain to occur compared with the results as shown in Figs. 37 and Fig. 38. The factors such as the elevated temperature and pores with different spacing and orientations are one of main reasons to affect on the local plastic distribution when the same applied stress.

## 4. Effects of testing conditions on fatigue life of cast magnesium alloys

Above mentioned fatigue life of typical cast magnesium alloys as shown in Fig. 32, these results involve that the effects of samples with a single side notch and with two pores at different orientations, spaces and vacuum states on the fatigue life can be not ignored. However, it is difficult to distinguish that the fatigue lives of cast magnesium alloys are how to be affected by these testing conditions even if we known that the fatigue damage mechanism and fatigue crack growth rate at the different conditions of cast magnesium alloys.

As majority cast magnesium alloys, the fatigue life can be divided into two parts: fatigue crack initial life and fatigue crack propagation life. The fatigue crack propagation life can be estimated by the fatigue crack growth rate such as Eq. (4-1) and the fatigue crack initial life has to be deduced based on the *S-N* curves of smooth samples or to do some fatigue tests with smooth samples at the minimal measurable length of fatigue crack initiation although it is rather difficult work. Therefore, many investigators focused on how to obtain the fatigue crack growth rate or *S-N* curves of materials at the different conditions by the experimental manners. Especially, when the fatigue crack propagation life of majority metals occupies about over than 70% of total fatigue life, it is very important for the fatigue crack propagation tests of materials, which brings significant advantages in estimation the fatigue life of cast magnesium alloys or a design of automotive and aircraft components. The fatigue crack propagation process is more important in service for most ductile materials because the fatigue life is mainly determined by the cycles spent, especially spent for the cycles of the small crack growth [49,50]. With the development of techniques in a microscopy, it is possible to accurately and directly measure the length of fatigue small cracks growing. Examples of this are *in-situ* observations with SEM that allow us to investigate the fatigue crack initiation and propagation of metals [51].

### 4.1 Effect of elevated temperatures on the fatigue life

Fig. 39 shows that the *S-N* curves of cast AM50 samples with a notch (the radius of notch is about 50 μm in these *S-N* curves) were carried out at different temperatures in a vacuum chamber. With increasing the elevated temperature, the fatigue life of the cast AM50 magnesium alloy decreased monotonously. This means the fatigue life of cast AM50 Mg alloy depended mainly on the fatigue crack growth rate at these temperatures. Viewing the fatigue failure patterns at elevated temperatures (more than 100 °C) when the applied stress

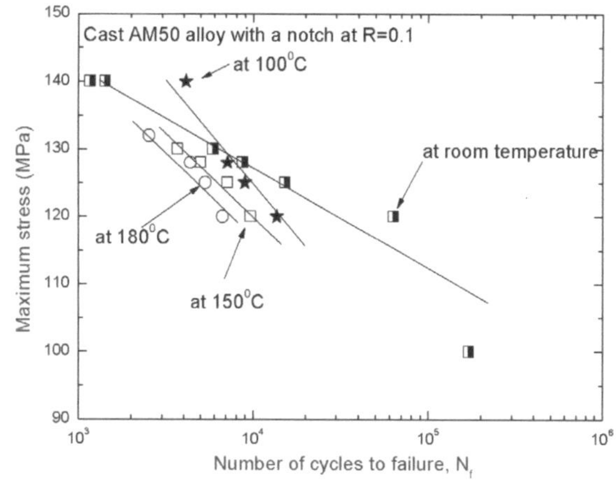


Fig. 39. *S-N* curves at different elevated temperatures of cast AM50 alloy

amplitude is a constant, it is found that the fatigue failure mechanism is mainly cleavage fracture of α-Mg grains and Mg-dendrites as shown in Figs. 40A (It can be classified as the Mode I crack.). This cleavage fracture is produced by repeated cycling on the crack tip plastic zone. Fatigue small cracks in vacuum chamber were accelerated with the increasing thermal-mechanical loading because the β-phase ($Mg_{17}Al_{12}$) in cast AM50 magnesium alloy began to soften at the high temperature. Compared with properties of fatigue cracks propagated at different temperatures in a vacuum chamber indicated that the propagation of fatigue small cracks mainly along the interdendrite region or along the boundary of α-Mg grains at the room temperature [14,15] and the fatigue small cracks propagated passing through inside of the Mg grains because these regions were soften at higher temperatures. As shown in Figs. 40B, there are some significant plastic slip vestiges which appeared on the up and down of the crack tip and lied at about 45° inclined with respect to the crack, so that the growth rate of small fatigue cracks at higher temperatures is larger than that at room temperature in a vacuum chamber. The fatigue small crack path is clear in Figs. 40C and Figs. 40D either cleft the α-Mg grains or passed along the boundary of the α-Mg grains. The former (transgranular fracture) was occurred more frequently than the latter at the high temperatures. Our observation results indicated that the small fatigue cracks preferentially grew in a Mode I direction. Microstructural small cracks, however, frequently deviate in order to avoid an obstacle or to link with a weakened material zone. Gall et al. have also

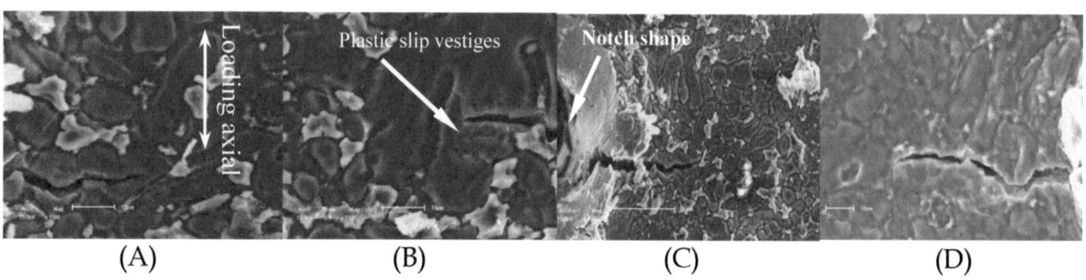


Fig. 40. SEM images of fatigue crack growth under different loadings at R=0.1. (A) $\sigma_{max}$=128 MPa, T=150°C, N=4109; (B) $\sigma_{max}$=125 MPa, T=150°C, N=5819; (C) $\sigma_{max}$=128 MPa, T=150°C, N=4260; (D) $\sigma_{max}$=120 MPa, T=180°C, N=12526.

obtained a similar conclusion for AM60B specimens tested in vapor environment [30]. Thus, the propagation rates of small fatigue cracks of Mg-Al alloys are strongly dependent on the temperatures.

#### 4.2 Effect of notch and oxidation on the fatigue life

To observe fatigue small crack initiation and propagation process, a notch in the edge of specimens were cut manually. As the typical case, the notch radius and depth of all notch specimens are approximately 50 μm, 100 μm, respectively. The fatigue tests with the smooth specimens of cast AM50 Mg alloy were also carried out to compare the difference between fatigue life of smooth and notched specimens either in the vacuum or in air. The corresponding *S-N* curves are shown in Fig. 41. It is clearly shown that the difference of these influences are greater under the higher stress level than under lower stress level, whether or not the environment is air or vacuum. On the other hand, this means that the effect of the notch on the fatigue life became rather obvious with increasing stress level. It is confirmed that the notch size mainly affects the fatigue initiation life of material.

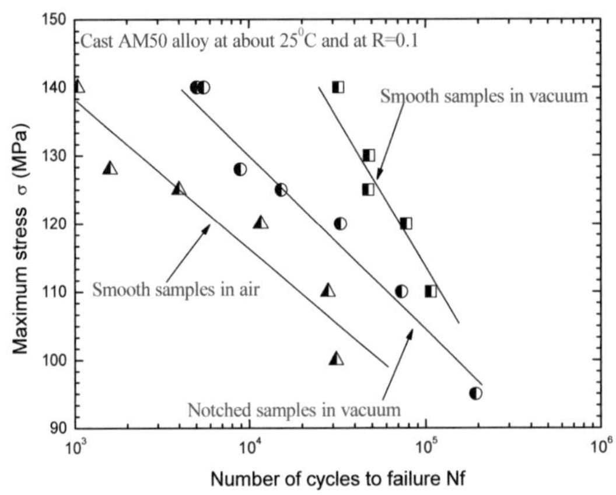


Fig. 41. Effect of notch and oxidation on *S-N* curves of cast AM50 alloy

The maximum von Mises stress value in the stress concentration region around the notch tip increases with the applied stress increasing. If the stress level in the stress concentration region reaches a certain magnitude such as over than 70% yield stress of cast magnesium alloy, the fatigue crack will occur in the stress concentration region after a several hundreds cycles. However, at the lower stress level such as less than 60% yield stress of cast AM50 alloy, the differences of the total fatigue lives of specimens with a notch and smooth are mainly contributed by the fatigue crack initiation life, which is controlled by stress concentration degree such as it refers to two parameters both different radiuses and depths of a notch. For example, this effect was reported that the number of cycles spent on fatigue crack initiation is less than one half of the total fatigue life for smooth specimens of the general metals [52]. Therefore, the effect of a notch on the fatigue life of cast magnesium alloys is similar to the reported results in Literature [52].

In addition, cast Mg alloys have generally a low oxidation resistance, thus there is a difference about fatigue life of cast magnesium alloys under different fatigue tests. For example, the two *S-N* curves of cast AM50 alloy indicates that the fatigue life in air is shorter than that in vacuum under the same stress amplitude as shown in Fig. 41. This means that the effect of air corrosion on the fatigue of this alloy is not ignored, especially in an absence of surface barriers on cast magnesium alloys, e.g. in the case without oxide films or anodized coating [30,52]. At the same time, the *S-N* curves indicated still that environmental effect both in air and vacuum on the fatigue life of cast magnesium alloy has the different trends at low and high stress levels. At the lower stress level, the effect of environmental effect both in air and vacuum becomes smaller and smaller, contrarily larger and larger at higher stress level for the smooth specimens.

### 4.3 Effect of cast magnesium alloys on the fatigue life

As cast magnesium alloys, the mechanical properties (especially the elongation (%) or fracture toughness) differ observably from each other, such as cast AM50 alloy and cast AZ91 alloy. Therefore, the fatigue crack initiation life of these materials is observably different. For example, Fig. 42 shows the fatigue curves of cast AM60 and AZ91 alloys at the different fatigue crack initiation and propagation tests. Due the fatigue tests are LCF, the fatigue life of cast magnesium alloys depends on the plastic strain energy so that the elongation (δ%) (or fracture toughness) of cast magnesium alloy dominates its fatigue life. The many fatigue cracks for quasi-brittle secondary phase in the cast magnesium alloy occur easily at their interface or boundary as shown in Figs. 43A. At the same time, the plastic

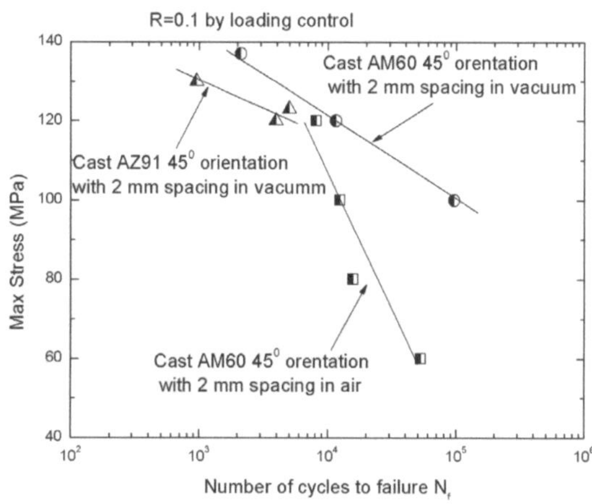


Fig. 42. Effect of environmental vacuum and magnesium alloys on *S-N* curves.

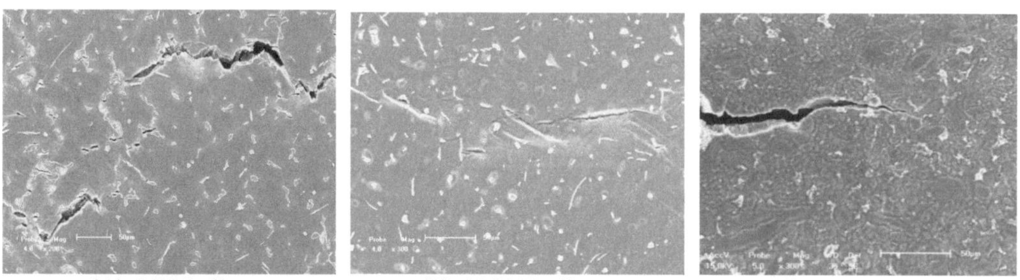

(A) AZ91 alloy (B) AM60B alloy (C) AM50 alloy

Fig. 43. Fatigue crack characteristics of different cast magnesium alloys

deformation will present at rather softness α-Mg phase near the fatigue crack as shown in Figs. 43B. Therefore, the fatigue propagation life of multi-cracks is rather shorter than that of single fatigue crack because there is combined near these cracks so that it causes to accelerate the fatigue crack growth rate of cast magnesium alloy. For the fatigue life of cast AM60 alloy in air and vacuum conditions, the differences both conditions can be not ignored as shown in Fig. 42.

## 5. Summary

Fatigue refers to mechanical failure (and the processes leading to it) when a cast magnesium alloy is subjected to a cyclical stress or strain amplitude that would not result in fracture during monotonic loading. This kind loading can take place in all classes of materials. However, it was not commonly thought to happen in a quasi-brittle cast magnesium alloy, for small cracks formed during fatigue of quasi-brittle cast magnesium alloy quickly lead to catastrophic fracture owing to the low fracture toughnesses of majority cast magnesium alloys. As a result, fatigue small crack initiation and propagation behavior, and relative several central issues dealing with the effects of some factors (temperature, notch and pores etc.) on the fatigue cracking behaviors and life of cast magnesium alloys (including AZ91D, AZ91, AM60B, AM60 and AM50) were investigated by the different experimental methods. The fatigue cracking behavior, fatigue life of cast magnesium alloys and the influencing factors were detailedly discussed in this chapter.

The fatigue crack initiation behavior (Stage I of the fatigue process) of cast magnesium alloys are formed initially at surface flaws and/or defects (including to the manual notch) that promote localized flow in their vicinity. Less frequently they nucleate at internal inhomogeneities (e.g., interdendritic pore, pores, secondary phase particles etc.) that likewise serve to promote stress concentrations region in the cast magnesium alloys surrounding them. These types of fatigue nucleation events are common to metals. This leads to the development of a feature that, at some stage, can be characterized as a Stage I crack as shown in Figs. 9B. In the initial growth of such, it propagates in a direction determined by slip crystallography, and this direction is not normal to the principal stress axis. Thus, Stage I crack propagation is defined by a flow, rather than a fracture, criterion. After such a crack has progressed a certain distance, it alters its direction so that it becomes normal to the principal stress axis or applied loading direction. At this point further advance depends on other factors similar to those applying to tensile fracture as shown at the 3.2 section in this chapter. However, the effects of rather smaller radius of notch and the different spacing and orientations of two pores can be not ignored.

Following this alteration of its course in micro scale, continued propagation of the crack takes place in an intermittent manner (the crack growth region, Stage II of fatigue). The crack growth rate, $d\ell / dN$ (the change in crack length $\ell$, with the number of cycles, $N$) is related to the "either a term of $\sigma_{\max}^{n}\ell$ at $R$=0.1 (or $\Delta\sigma^{n}\ell$ at $R$=-1). The higher the value of $\sigma_{\max}^{n}\ell$ or $\Delta K$, the greater of $d\ell / dN$ is, and $d\ell / dN$ of a given cast magnesium is uniquely related to the term of $\sigma_{\max}^{n}\ell$ or $\Delta K$. For most cast magnesium alloys there exists a threshold cyclical stress intensity, $\Delta K_{th}$ below which fatigue cracks will not propagate. Sometimes, if the fatigue crack growth rate is rather lower than the given value, such as the rate of about $10^{-10}$ m/cycle, it can be defined as the threshold value of fatigue crack growth rate. Using the cast magnesium alloys at $\Delta K$ values less than $\Delta K_{th}$ constitutes "fail-safe" fatigue-fracture design. Unfortunately, $\Delta K_{th}$ is usually quite low, so that use of it in design constitutes inefficient cast magnesium alloys utilization. Consequently, fatigue design philosophy assumes the presence of fatigue cracks that will grow during service, but also employs empirical guidelines coupled with timely service inspection to see to it that such cracks do not progress to the extent that cast magnesium alloys failure occur during the intended service life. As understanding the fatigue cracking mechanism and fatigue crack growth rate of cast magnesium alloys at the certain condition, the residual life of cast magnesium alloy can be estimated by continuously detecting the measurable fatigue crack propagation length.

For most cast magnesium alloys are best suited to LCF application. This is related to the fact that the Stage II growth occupies most of the material's life during low-cycle fatigue. It is clear that hard, quasi-brittle cast magnesium alloys do not resist crack advance.

Temperature has an influence on fatigue response of cast magnesium alloys. This is important for evaluated temperature of cast magnesium alloys influence on the fatigue crack propagation mechanism. For example, the fatigue crack initiation of cast AM50 alloy occurred at the root of notch (when the radius of notch is less than the 50 μm) but the early stage of crack propagation is along either the boundary of α-Mg grain or to cleave the α-Mg grain in front to the crack tip. The fatigue crack propagation mechanism in the microscopically zone is analogous to the quasi-brittle or quasi-ductile (intervenient brittle and ductile) fracture mechanism of engineering alloys. At RT, we did not find that the α-Mg grain of cast AM50 alloy was cleaved. When the elevated temperature is over than 100 °C, the fatigue crack propagation is also either along the boundary of α-Mg grain or to cleave the α-Mg grain. In addition, the fatigue crack propagation mechanism of cast AM50 alloy at the elevated temperature indicated that the branch fatigue crack was found as shown in Figs. 13B. This means that the fracture mechanism of cast magnesium alloy at the elevated temperature which is Mode I and Mode II differs obviously from that which is only Mode I at room temperature in microscopically zone. Corresponding effect of the elevated temperatures on the fatigue cracking mechanism of cast magnesium alloys has analogous to composite fractures of Mode I/II. Therefore, the effect of the elevated temperature on the fatigue crack propagation mechanism of cast magnesium alloys can be not ignored. This is because the elevated temperature easily causes the β-$Mg_{17}Al_{12}$ becomes a softness so that the deformation mismatch between the α-Mg grain and β-$Mg_{17}Al_{12}$ phase becomes weak at the elevated temperatures. This is a competitive result of the interface strength and the fracture strength of α-Mg grain.

Small pores or radius of notch has an influence on the fatigue crack propagation behavior of cast magnesium alloys, which involves the FE calculation for the stress/strain distribution around each pore. All calculated von Mises stresses can be scaled with respect to magnitude of the applied stress or transmit strain based on the constitutive equations of cast magnesium alloys. The stress distribution will be illustrated by the principal stress and the von Mises effective stress on crack propagation plane around the pore. It is approximately equivalent to the axial stress and vanishes at the pole and takes its largest values at the overlap region of plastic deformation of cast magnesium alloy.

As the effects of spacing and orientations with two pores on the fatigue cracking behavior of cast magnesium alloys, these experimental results indicated that there is a critical spacing value of two pores with the 90° or 45° orientations whether to tack place the coalescence of crack at one pore with another one, which is about 2.0 mm either 90° or 45° orientations with two pores. The effect of different orientations on the fracture strength and fracture toughness of cast AM60 alloy under the static tensile loading is obvious. That is, the fracture strength and fracture toughness of a specimen with 90° orientations are lower than that of a

specimen with 45° orientations. This means that the former damages or fractures prior to the latter. These experimental conclusions can be validated and explained by the difference of stress concentration area fraction as shown in Fig. 36.
Cast magnesium alloys have generally a low oxidation resistance, thus there is a difference about fatigue life of smooth and notch specimens under different fatigue tests. For example, the two S-N curves of cast AM50 Mg alloy indicates that the fatigue life in air is shorter than that in vacuum under the same stress amplitude as shown in Fig. 41. This means that the effect of air corrosion on the fatigue life of this alloy is not to be ignored, especially in an absence of surface barriers on cast magnesium alloys, e.g. in the case without oxide films or anodized coating [30,52]. At the same time, the S-N curves indicated still that environmental effect both in air and vacuum on the fatigue life of cast magnesium alloy has the different trends at low and high stress levels. At the lower stress level, the effect of environmental effect both in air and vacuum states becomes smaller and smaller, contrarily larger and larger at the higher stress level for the smooth specimens as shown in Fig. 41.

## 6. Acknowledgement

The author would like to thank Prof. Fan Jing-Hong, Prof. Tang Bin and Dr. Wu Bi-Sheng to be cooperated in past decades. At the same time, the author would like to thank the projects (Grants No. 50571047, 11072124) supported by NSFC, National Basic Research Program of China through Grants No. 2007CB936803, 2010CB631006 and by State Key Lab of Advanced Metals and Materials in Uni Sci Tech Beijing (2010ZD-04).

## 7. References

[1] Mordike BL, Ebert T. Magnesium properties-applications-potential. Mater Sci Eng A 2001; 302: 37-45.
[2] Sakkinen DJ. Physical metallurgy of magnesium die cast alloys. SAE Technical Paper 940779, Detroit (MI): SAE; 1994.
[3] Mayer H, Papakyriacou M, Zettle B, Stanzl-Tschegg SE. Influence of porosity on the fatigue limit of die cast magnesium and aluminium alloys. Int J Fatigue 2003; 25: 245-256.
[4] Kadiri Haitham EI, Xue YB, Horstemeyer MF, Jordon JB, Wang PT. Identification and modeling of fatigue crack growth mechanisms in a die-cast AM50 magnesium alloy. Acta Mater 2006; 54: 5061-5076.
[5] Horstemeyer MF, Yang N, Gall K, McDowell DL, Fan J, Gullett PM. High cycle fatigue of a die cast AZ91E-T4 magnesium alloy. Acta Mater 2004; 52: 1327-1336.
[6] Gall K, Biallas G, Maiers HJ, Horstemeyer MF, McDowell DL. Environmentally influenced microstructurally small fatigue crack growth in cast magnesium. Mater Sci Eng A 2005; 396: 143-154
[7] Shih TS, Liu WS, Chen YJ. Fatigue of as-extruded AZ61A magnesium alloy. Mater Sci Eng A 2002; 325: 152-162.
[8] Biallas G and Maier HJ. In-situ fatigue in an environmental scanning electron microscope-potential and current limitations. Int J Fatigue 2007; 29: 1413-1425.
[9] Tang B, Li SS, Wang XS, Zeng DB. An investigation on hot-crack mechanism of Ca addition into AZ91D alloy. J Mate Sci 2005; 40(11): 2931-2936.
[10] Tang B, Wang XS, Li SS, Zeng DB. Effects of Ca combined with Sr additions on microstructure and mechanical properties of AZ91D magnesium alloy. J Mate Sci & Tech 2005; 21(5):574-578.

[11] Tang B, Li SS, Wang XS, Zeng DB, Wu R. Effect of Ca/Sr composite addition into AZ91D alloy on hot-crack mechanism. Scripta Mater 2005; 53(9):1077-1082.
[12] Kim KC, Nam SW. Effects of Mn-dispersoids on the fatigue mechanism in an Al-Zn-Mg alloy. Mater Sci Eng A 1998; 244(2): 257-262.
[13] Wang XS, Jin L, Li Y, Guo XW. Effect of equal channel angular extrusion process on deformation behaviors of Mg-3Al-Zn alloy. Mater Lett 2008; 62:1856-1858.
[14] Wang XS, Lu X, Wang HD. Investigation of surface fatigue microcrack growth behavior of cast Mg-Al alloy. Mater Sci Eng A 2004; 364: 11-16.
[15] Wang XS, Fan JH. SEM online investigation of fatigue crack initiation and propagation in notched cast magnesium specimens. J Mate Sci 2004; 39(7): 2617-2620.
[16] Wang XS, Fan JH. An evaluation the growth rate of small fatigue cracks in cast AM50 magnesium alloy at different temperature in vacuum environment. Int J Fatigue 2006; 28(1): 79-86.
[17] Wang XS, Liang F, Fan JH, Zhang FH. Investigations on low-cycle fatigue small crack initiation and propagation mechanism of cast magnesium alloys based on in-situ observation with SEM. Philos Mag 2006; 86(11):1581-1596.
[18] Wang XS, Fan JH. Growth rate of small fatigue cracks of magnesium alloy at different conditions. Mater Sci Forum 2007; 546-549: 77-80.
[19] Fouret C, Degallaix S. Experimental and numerical study of the low-cycle fatigue behavior of a cast metal matrix composite Al-SiCp. Int J Fatigue 2002; 24(2-4): 223-232.
[20] Blochwitz C, Tirschler W. In-situ scanning electron microscope observations of the deformation behaviour of short cracks. Mater Sci Eng A 2000; 276(1-2):273-276.
[21] Andersson H, Persson C. In-situ SEM study of fatigue crack growth behaviour in IN718. Int J Fatigue 2004; 26(3):211-219.
[22] Marx M, Vehoff H. Propagation of microcracks in single crystalline nickel-based superalloys: size effects on the crack opening. Mater Sci Eng A 2004; 387-389: 511-515.
[23] Gilbert CJ, Ritchie RO. On the quantification of bridging tractions during subcritical crack growth under monotonic and cyclic fatigue loading in a grain-bridging silicon carbide ceramic. Acta Mater 1998; 46(2): 609-616.
[24] Tanaka Y, Deng ZY, Liu YF, Masuda C. In situ observation on fatigue crack growth in SCS-6/Ti-15-3 composite at elevated temperatures. Acta Mater 2003; 51(20): 6329-6340.
[25] Weidner A, Tirschler W, Blochwitz C. Overstraining effects on the crack-opening displacement of microstructurally short cracks. Mater Sci Eng A 2005; 390(1-2): 414-422.
[26] Hadianfard MJ, Mai YW. In situ SEM studies on the effects of particulate reinforcement on fatigue crack growth mechanism of aluminium-based metal-matrix composite. J Mater Sci 1995; 30(21): 5335-5346.
[27] Mutoh Y, Zhu SJ, Hansson T, Kurai S, Mizuhara Y. Effect of microstructure on fatigue crack growth in TiAl intermetallics at elevated temperature. Mater Sci Eng A 2002; 323(1-2): 62-69.
[28] Chan KS, Jones P, Wang Q. Fatigue crack growth and fracture paths in sand cast B319 and A356 aluminum alloys. Mater Sci Eng A 2003; 341(1-2): 18-34.
[29] Tokaji K, Kamakura M, Ishiizumi Y, Hasegawa N. Fatigue behavior and fracture mechanism of rolled AZ31 magnesium alloy. Int J Fatigue 2004; 26: 1217-1224.
[30] Gall K, Biallas G, Maiers HJ, Gullett P, Horstemeyer MF, McDowell DL, et al.. In situ observations of high cycle fatigue mechanisms in cast AM60B magnesium in vacuum and water vapor environments. Int J Fatigue 2004; 26: 59-70.
[31] Wang XS, Fan JH, Wu BS, Li Y. Effects of spacing and alignment pores on fatigue crack behaviors of cast magnesium alloys. Adv Fract & Mat Behavior 2008; 33-37:13-18.

[32] Wang RM, Eliezer A, Gutman EM. An investigation on the microstructure of an AM50 magnesium alloy. Mater Sci Eng A 2003; 355: 201-207.
[33] Horstemeyer MF, Yang N, Gall K, McDowell DL, Fan J, Gullett PM. Fat Fract Eng Mater Struct 2002; 25; 1045-1056.
[34] Padfield TV, Handbook of metallography and microstructures, vol. 9. Materials Park (OH): ASM; 2004.
[35] Lu YZ, Wang QD, Zeng XQ, Ding WJ, Zhao CQ, Zhu YP. Effects of rare earths on the microstructure, properties and fracture behavior of Mg-Al alloys. Mater Sci Eng A 2000; 278: 66-76.
[36] Moreno IP, Nandy TK, Jones JW et al. Microstructural characterization of a die-cast magnesium rare earth alloy. Scripta Mater 2001; 45: 1423-1429.
[37] Zheng WC, Li SS, Tang B, Zeng DB, Guo XT. Effect of rare earths on hot cracking resistant property of Mg-Al alloys. J Rare Earth 2006; 24(3): 346-351.
[38] Li SS, Zheng WC, Tang B, Zeng DB, Guo XT. Grain coarsening behavior of Mg-Al alloys with mischmetal addition. J Rare Earth 2007; 25(2): 227-232.
[39] He SM, Peng LM, Zeng XQ, Liu LP, Wang QD, Ding WJ, Zhu YP. Effects of variable La/Ce ratio on microstructure and mechanical properties of Mg-5Al-0.3Mn-1RE alloys. Mater Sci Forum 2005; 488-489: 231-234.
[40] Paris PC, Erdogan F. A critical analysis of crack propagation laws. Trans ASME J Basic Eng 1963; 85: 528-534.
[41] ASTM: E647. Standard test method for measurement of fatigue crack growth rates. ASTM standards. The American Society for Testings and Materials 03.01 1998; p562-598.
[42] Ramberg W, Osgood WR. Description of stress-strain curves by three parameters. *National Advisory Committee for Aeronautics -- Technical Notes* 1943; p13.
[43] Wang XS, Kawagoishi N. Simple predicting method for fatigue crack growth rate based on tensile strength of carbon steel. J Iron & Steel Res Int 2003; 10(2): 58-62.
[44] Wang XS, Xu Y, Xu XQ. Direct observations of microcracking in the fuel plate using the scanning electron microscope. J App Comp Mater 2004; 11(3):145-154.
[45] Wang XS, Wu BS, Wang QY. Online SEM investigation of micro fracture characteristics of concretes at various temperature values, Cement Concrete Res 2005; 35(7): 1385-1390.
[46] Wang XS, Liang F, Zeng YP, Xie XS. SEM in-situ observations to investigate the effects of inclusions on the low cyclic fatigue crack initiation and propagation of super strength steel, Acta Metall Sin 2005; 41(12):1272-1276.
[47] Wang XS, Zhang LN, Zeng YP, Xie XS. SEM in-situ investigation on fatigue cracking behavior of P/M Rene95 alloy with surface inclusions. J Uni Sci Tech Beijing-Mater 2006;13(3): 244-249.
[48] Wang XS, Yan CK, Li Y, Xu YB, Meng XK, Wu BS. SEM in-situ study on failure of nanocrystal metallic thin films and substrate structure under three point bending. Int J Fracture 2008; 151: 269-279.
[49] Rice JR, ASTM STP 1973; 536: 231-245.
[50] Wang XS, Xu Y. Experiments, characterizations and analysis of a dispersion U3Si2-Al fuel plate with sandwich structure. J Nucl Mater 2004; 328(2-3): 243-248.
[51] Kawagoishi N, Wang XS, Nisitani H, Goto M, Kondo E, Prediction of Fatigue Life Based on Small Crack Growth Law. Trans JPn Soc Mech Eng A 1997; 63(613): 1867-1873.
[52] Gutman EM, Eliezer A, Unigovski YE, Abramov E., Mechanoelectrochemical behavior and creep corrosion of magnesium alloys. Mater Sci Eng A 2001; 302: 63-67.

# 5

# Biocompatible Magnesium Alloys as Degradable Implant Materials - Machining Induced Surface and Subsurface Properties and Implant Performance

Berend Denkena[1], Arne Lucas[1], Fritz Thorey[2], Hazibullah Waizy[2], Nina Angrisani[3] and Andrea Meyer-Lindenberg[3]
[1]*Institute of Production Engineering and Machine Tools, Leibniz Universität Hannover,*
[2]*Department of Orthopaedic Surgery, Hannover Medical School,*
[3]*Small Animal Clinic, University of Veterinary Medicine Hannover,*
*Germany*

## 1. Introduction

Annually several million people suffer bone fractures caused by accidents or diseases. Many of those fractures are too complex for an external medical treatment but have to be surgically fixated by internal bone implants. Traditional methods of osteosynthesis or osteotomy use permanent metal implants e.g. bone screws and bone plates made of steel or titanium alloys, but permanent metal implants have to be excised. Especially young patients in growth require the implant removal. Usually, metal implants should be removed latest one or two years after the first surgery. Biodegradable implants, which dissolve in the human organism, therefore represent an appropriate solution. Here magnesium as a degradable implant material provides both biocompatibility and sufficient mechanical properties. Studies have shown that magnesium, which is an essential element of the human organism, is suitable as a degradable biomaterial for use in medical implants. Also in relation to biomechanical requirements the suitability of magnesium as an implant material in has already been proven. Due to mechanical properties superior to the bone, magnesium implants show biomechanical properties comparable to conventional steel implants in standardized biomechanical tests. Also an influence of surface properties on the degradation behavior is already known. Interdisciplinary research between medical and engineering sciences focuses on the development of degradable magnesium implants for osteosynthesis. The removal of those implants after convalescence of the fractured bone is no longer necessary, resulting in a considerable benefit for patients and the public health care system.

The aim is to preset individual degradation kinetics for implants made of biocompatible magnesium alloys through adjusting surface and subsurface properties of the magnesium implants for specific indications. Adapted machining processes exhibit possibilities to adjust those properties. The idea of implants specifically adapted to their desired function and functional period within a possible range of surface or subsurface modifications is illustrated schematically in Figure 1for different implant designs (V1 and V2). The healing process of a bone is illustrated schematically and opposed to the expected corresponding degradation process of a resorbable implant. Furthermore, possible processes for mechanical machining of the workpieces and modification of the surface and subsurface are illustrated. By means of these processes specific surface and subsurface properties can be adjusted, which, depending on the alloy, can achieve the modification bandwidth illustrated in Figure 1. By means of this modification bandwidth an implantation behavior of the implants during the production process can be achieved specific for each indication.

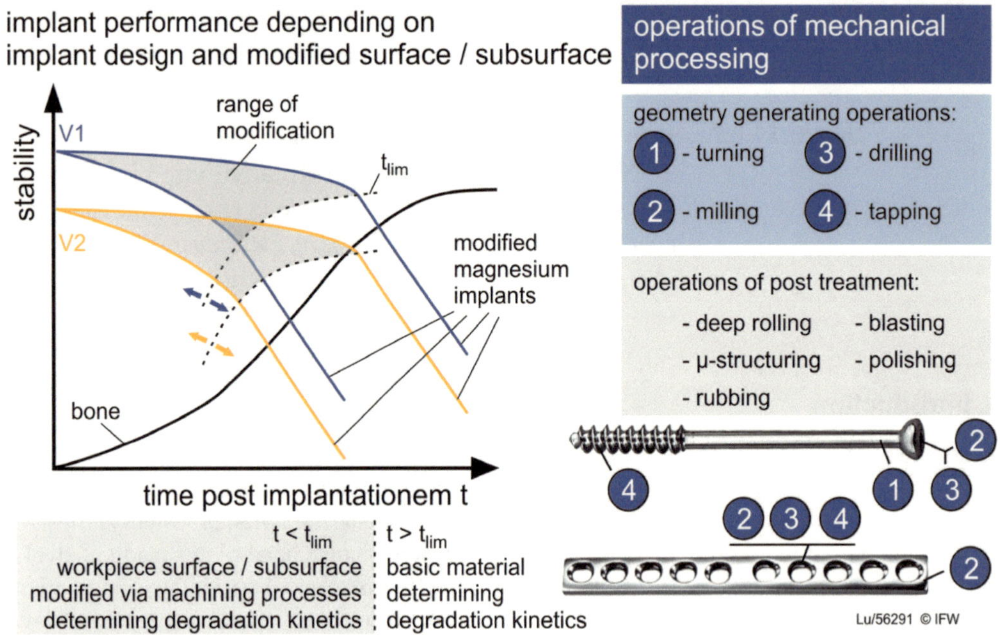


Fig. 1. Scheme of implants adapted to function and applicable machining processes

To adjust specific implant properties, the influence of the mechanical processing on surface and subsurface layers on selected biocompatible magnesium alloys has to be analyzed. Different operations of processing and post-processing magnesium workpieces enable the adjustment of distinct surface and subsurface properties. These properties are then utilized to adjust the corrosion resistance and therefore the degradation kinetics of the implant in the organism. Results of in-vitro an in-vivo experiments validate the strong relation between workpiece properties and corrosion behavior respectively degradation kinetics.

The influence of different processes for the mechanical processing as well as the process input values and process parameters or tool geometries on the workpiece properties and the resulting degradation behavior are investigated on newly developed biocompatible magnesium alloys. Biomechanical characteristics of the samples produced are then analyzed and the influence of the corrosion attack on the biomechanical characteristics of the components is determined in different states of corrosion. In clinical in-vivo trials the degradation kinetics, the profile of biomechanical functions and the osseointegration of various implant demonstrators are investigated. The combination of the results of the in-vivo and in-vitro studies serves as a basis for the derivation of biomechanically relevant parameters and therefore for a definition of the medical requirements specifically for resorbable osteosynthesis implants. In the future, this profile of requirements will determine the implant design, the machining processes and the corresponding process parameters and therefore enables a conception of implants adapted to various indications.

## 2. State of science and technology

In Orthopaedics, implants such as screws, plates, intramedullary nails etc. are utilized for fracture fixation (osteosynthesis). Theses implants, which currently are mainly produced out of steel and titanium materials, are used for the support or fixation of traumatized bones and therefore, in most cases, only have a temporary function until the healing of the bone is completed. The formerly necessary revision surgery, which implies a certain risk for the patient as well as additional costs, can be avoided by utilizing degradable implants. In Orthopaedics and fracture fixation temporary bone implants therefore shall replace more and more permanent osteosynthetic materials, especially in the field of small volume implants (Weiler et al., 2000). These implants temporarily take over the function of the bone for a mechanical stabilization until after gradual degradation they are being replaced by newly built up bone tissue. The temporary implant materials utilized should guarantee a suitable mechanical stability and at the same time be completely degradable and should correspond to the supporting bone (cortical bone) regarding mechanical characteristics to avoid stress shielding (Daniels et al., 1990; Weiler et al., 1998, 2000). Until now, temporary implants are mainly produced out of PLA (polylactic acid, D and L isomers), PGA (polyglycolic acid), TMC (trimethylene carbonate), copolymers with varying substantial ratios, PDS (polydiaxonone) as well as composite materials (calcium phosphate in combination with polymers) (Anker et al., 2005; Ignatius et al., 2001a, 2001b; Sakar et al., 1999; Taylor et al., 1994). The field of application of these materials is however limited due to their relatively low mechanical strength. Therefore they can only be utilized in areas with low tensile, shear and compression forces (Wittenberg et al., 1991). Magnesium-based alloys specifically developed for medical applications offer significantly higher ultimate strengths, a higher Young's modulus and a proven biocompatibility (Niemeyer et al., 2005; Wintermantel & Ha, 1996).

Chip removing machining processes are generally used for shaping of medical implants because they guarantee a defined influence of the surface and subsurface, in contrast to other techniques such as forming. The flexibility of the design of the sample implants is an additional advantage of cutting machining processes. Regarding the resulting subsurface properties due to the different chip removing machining techniques there are still numerous questions to be answered. Due to the degradation of metallic implants mainly resulting from friction and corrosion, achieving defined subsurface properties in the production of function adapted medical implants is decisive. In the past few years, research on the influence of different surface machining techniques for implants on reactions of the organism has mainly been carried out in the field of dentofacial surgery. Research was mainly done with non-resorbable materials, especially titanium and titanium alloys. The surface topography was the center of attention (Gotz et al., 2004; Nasatzky et al., 2003; Schneider et al., 2003; Shalabi et al., 2006). Further studies showed that of the otherwise identical implants the ones with a thicker oxide layer created by heat treatment of the material lead to an earlier strong implant-bone compound (four weeks after implantation) but also revealed that this difference is being neutralized after a certain time, here 12 weeks after the implantation (Kim et al., 2003). Research on the influence of different shot peening methods during the

production of roughed surface implants showed, that shot peening of titanium implants with bioceramics in compared to $Al_2O_3$ led to increased bone formation at the implant (Mueller et al., 2003). Though, these studies were carried out on permanent implant materials and therefore, the influence on surface, subsurface and implantation behavior illustrated is not transferrable to the resorbable implant material magnesium.
Besides the machining processes also mechanical surface treatment by means of hardening and softening or micro and macro residual stress can influence the subsurface conditions. Additionally, depending on the process and the parameters, changes of the surface topography can be achieved. These effects can increase the fatigue strength and therefore influence the behavior of the component (Broszeit & Adelmann, 1996; Scholtes & Vöhringer, 1996; Scholtes & Zinn, 1997). Residual compressive stress within the subsurface reduces the intensity of tension at the crack tips and reduces the crack formation and component distortion (Kloos et al., 1988; Nowag et al., 2007; Nüstedt et al., 1996; Brinksmeier et al., 2007). On rotationally symmetric components the supporting proportion could be increased by means of deep rolling. This led to improved operation characteristics as well as a significant improvement of fatigue strength and corrosion resistance. The corrosion speed and therefore the degradation behavior of magnesium-based alloys, besides its composition, especially depend on the surface and subsurface properties after machining. The properties can be adjusted by mechanical processing and it is known that these alterations have an influence on the component properties (Denkena et al., 2006a, 2008; Tomac & Tonnessen, 1991; Tönshoff et al., 2007; Winkler, 2007). First investigations with biocompatible magnesium alloys already show promising results regarding biomechanical properties and the potential to adjust the corrosion behavior (Denkena et al., 2006b, Denkena & Lucas, 2007). However, the fundamental relations between the mechanical processing and the resulting surface and subsurface properties of biocompatible magnesium alloys have not yet been investigated. Furthermore, the mechanisms of action of these properties regarding the corrosion behavior and therefore the kinetics of degradation of implants are still unknown. This knowledge is necessary to enable the function adjusted processing of small magnesium implants.
The subsequent biomechanical investigation of resorbable screws, plates etc. regarding biomechanical characteristics for the application in orthopedic surgery, traumatology and maxillofacial surgery is oriented towards the testing of rigid materials. This currently implies the testing of compression forces, pull-out forces, shear strength and bending fatigue strength in corresponding test setups (Carano et al., 2005; DeCoster et al., 1990; Ignatius et al., 2005; Inceoglu et al. 2004; Seebeck et al., 2005; White et al., 2006). The application of synthetic bone out of polyurethane foams according to ASTM-standard F543 is usual (ASTM-Standard F543, 2002). The investigations are carried out on identical testing stands like e. g. the testing device "MTS 858 MiniBionix".

## 3. Results

Firstly, the relations between the particular process input values of the investigated mechanical processes for the production of implants such as cutting parameters, tool geometries or process control variables, regarding the resulting material removal processes and process forces as well as their influence on the surface topography and the subsurface are presented. Based on these findings, the influence of the modified surface and subsurface on the in vitro corrosion behavior is being determined in a time-resolved manner. Parallel the loss of function of implant demonstrators is being correlated to the corrosive weight loss in biomechanical investigations. The in-vivo trials on corrosion behavior and biomechanics allow the comparison of the time scales of in-vivo and in-vitro investigations. Considering the different time scales assumptions about the influence of mechanical processing on the degradation kinetics and the expected biomechanical loss of function can be made. The results of the different topic areas "Mechanical Processing", "Clinical in vivo trials" and "Biomechanics" are outlined in the following.

### 3.1 Mechanical processing

The focus of research in the field "Mechanical Processing" is placed on the clarification of the connections between mechanical processing and the in-vitro corrosion behavior. Therefore, the influence of machining processes with different chipping and non-chipping techniques and process kinematics on resulting surface and subsurface properties is analyzed and correlations with the in-vitro corrosion kinetics for different magnesium alloys are developed. All experiments described are carried out on alloys, which were developed for biomedical applications and provided as extruded profiles by the Institute of Material Science (IW) of the Leibniz Universität Hannover (Seitz et al., 2010). Subsequent in-vitro investigations of specifically modified workpieces provide information on the influence of different subsurface conditions on the degradation behavior. In these experiments the corrosion behavior is analyzed quantifiable and time-resolved.
As a result, the parameters, which generate high passive forces (rolling pressure, cutting edge rounding), prove to be most significant. This passive force acts normally to the workpiece surface and is therefore significantly involved in the work hardening of the subsurface. The modification bandwidth of the corrosion behavior is significantly influenced by the surface properties (micro topography of the workpiece surface) and the residual compressive stress profile in the subsurface. The possibility for the modification of the corrosion kinetics by means of the subsurface condition also strongly depends on the alloy utilized. With the described machining processes and procedures also implant demonstrators were produced for biomechanical testing and clinical in vivo trials.

#### 3.1.1 Characterization of the workpiece subsurface after machining

The comprehensive characterization of the workpiece subsurface after machining provides information on the influence of cutting conditions and process parameters on the machining result. The interactions between different machining techniques, cutting materials, tool geometries and process parameters are investigated regarding plastic deformation of the surface, burr formation, surface quality and modification of the subsurface. Therefore the surface topography is analyzed by tactile and optical methods. Furthermore, a specially equipped SEM is used, which also serves for the EDX analysis of material components and corrosion products. In cooperation with the IW micro computer tomographic (μ-CT) analysis of corroded workpieces are carried out. In the characterization of the machined samples the depth-influence of the machining processes were especially focused on. Therefore, the subsurface is characterized by means of depth resolved micro hardness profiles, structural analysis and residual stress depth profiles. To be able to detect the residual stress condition, residual stress depth profiles for each of the differently machined samples are generated with an X-ray diffractometer using the $\sin^2$-$\psi$-method.

#### 3.1.2 Studies on the machining influence under turning kinematics

On rotationally symmetric workpieces machining processes with rotational kinematics are applied. Deep rolling is used for the post processing of previously turned samples.

Experiments to compare the capability of influencing properties of different materials were carried out on the alloys MgCa0.8 and MgCa3.0. Therefore, samples of both materials are machined by means of turning and deep rolling at identical process parameters. Rolling forces of 50 N and 200 N are adjusted via the hydraulic pressure of the hydrostatic deep rolling tool Ecoroll HG6, in which a hard material ball rolls over the workpiece surface.

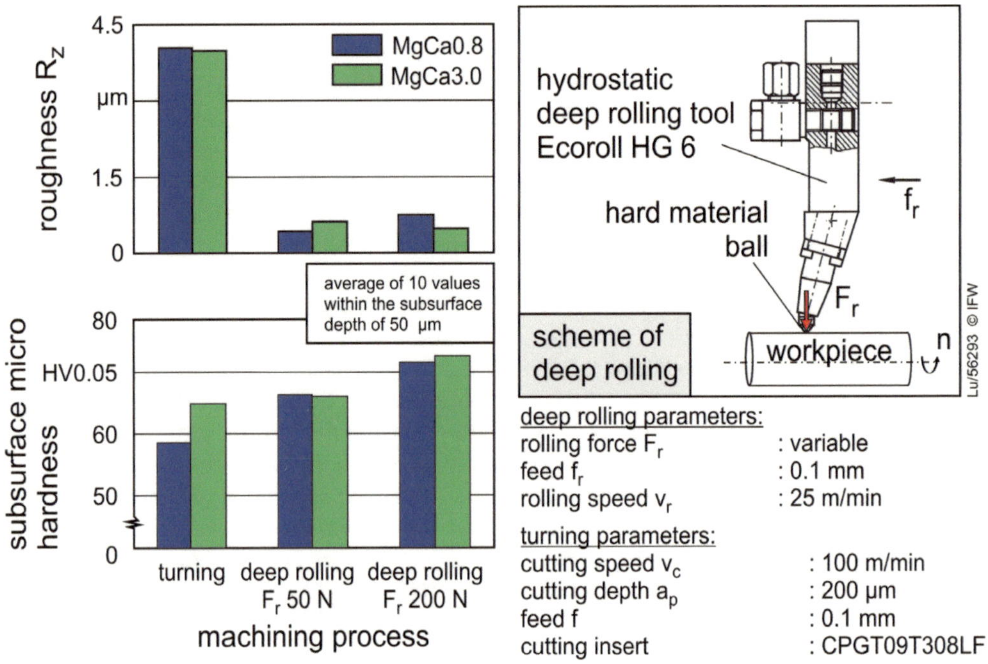


Fig. 2. Influence on surface and subsurface after turning and deep rolling

Some of the surface and subsurface properties resulting from processing are shown in Figure 2 and Figure 3 show. The surface quality is significantly improved by the deep rolling process whereas no differences concerning the alloy can be detected. In addition to this, an increase in hardness of the subsurface is achieved by increasing the rolling pressure. This becomes more obvious in the MgCa0.8 alloy, which is of lower basic hardness. Deep rolling as a post processing method provides the possibility to induce residual compressive stress in the subsurface.

Figure 3 shows the induced residual stress for MgCa0.8 and MgCa3.0. After turning, only a minor influence on the residual stress very closely under the surface can be detected in both alloys. After deep rolling, though, a strong shift of the residual stress profile in the compressive direction can be detected. By means of a rolling force of 50 N a higher residual stress maximum of approx. 150 MPa at a low depth is achieved whereas it is only 120 MPa for a rolling force of 200 N. A rolling force of 200 N, though, achieves an influenced depth of approx. 800 µm.

The results for different rolling forces correspond with the fundamental physical correlations of the Hertzian contact stress. Both alloys show similar subsurface properties and therefore similar possibilities of modification from the mechanical machining processes applied. The resulting corrosion characteristics will be described in a chapter below.

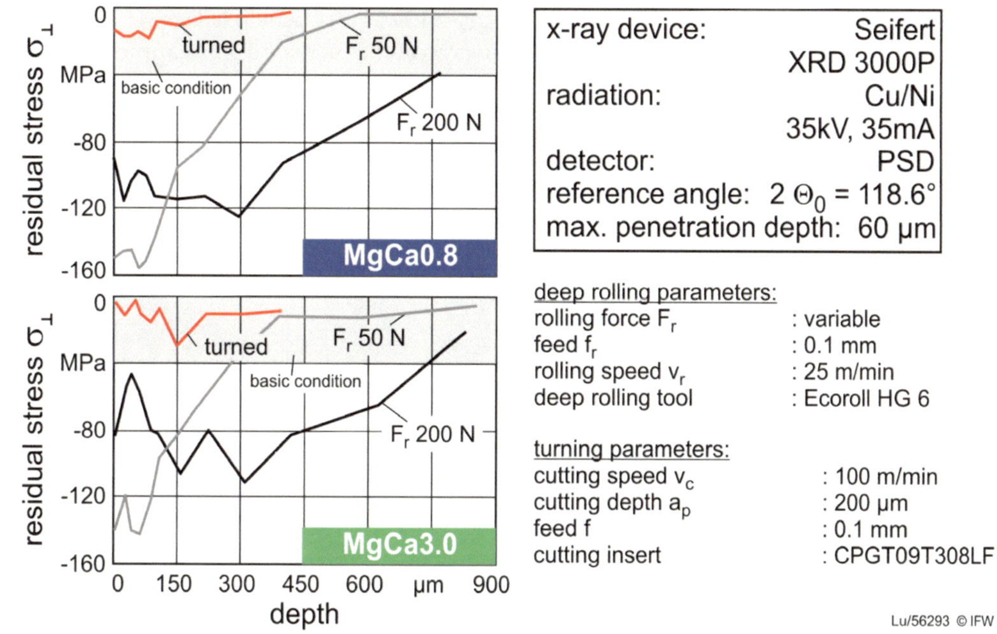


Fig. 3. Manipulation of subsurface after turning and deep rolling

### 3.1.3 Studies on the machining influence under milling kinematics

To be able to analyze in the influence of milling processes on surface and subsurface properties, magnesium cylinders of the alloy MgCa0.8 are face-milled. Therefore, process parameters are systematically varied and different tools are used. Following, the results from the investigations with milling tools, that were utilized with sharp edged ($r_\beta < 10$ µm) and modified cutting edges are shown. To modify the tools, the edges were prepared with cutting edge radii of $r_\beta = 100$ µm and $r_\beta = 200$ µm by means of brushing. Figure 4 shows the SEM-pictures of the tool variants, an icon of the face-milling process and the resulting surface qualities after milling.

At low cutting speeds, the surfaces created by honed tools are significantly rougher than on the samples machined with sharp cutting edges. The rounding leads to an undefined cutting process and therefore to increased formation of scratches and flakes. This effect disappears at higher cutting speeds and therefore increased thermal energy in the deformation zone in front of the cutting edge. Rising temperatures and increasing ductility of the material obviously lead to a smearing of the scratches and flakes and therefore a smoothing of the surface when using honed tools. The surface qualities resulting are similar to those of sharp tools.

By using honed cutting edges in face milling also a significant influence on the subsurface is achieved. Figure 5 shows that subsurface properties regarding the residual stress profile are generated, that are similar to those after deep rolling. While sharp tools hardly cause changes in the subsurface, honed cutting edges induce residual compressive stresses up to approx. 400 µm of depth. The penetration depth can be increased by advanced rounding. A residual stress maximum of up to – 120 MPa can be achieved for both roundings. The maximum for the rounding of $r_\beta = 200$ µm can be located at a depth of approx. 100 µm and therefore a little deeper than for the rounding of $r_\beta$ = 100 µm for which is located at a depth of approx. 70 µm. The penetration depth of influence can be increased from approx. 200 µm to approx. 400 µm be means of increased cutting edge rounding.

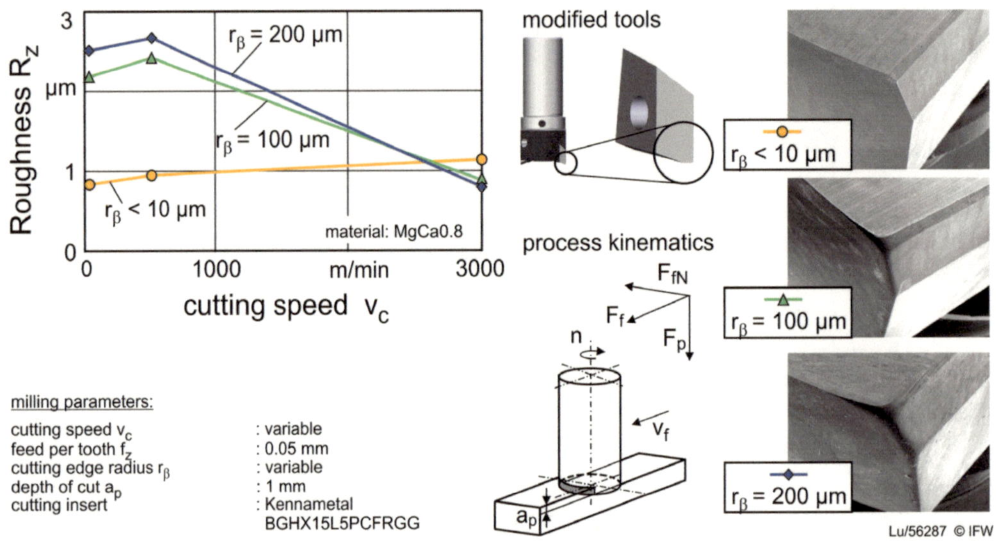


Fig. 4. Modified tools and surface qualities after milling

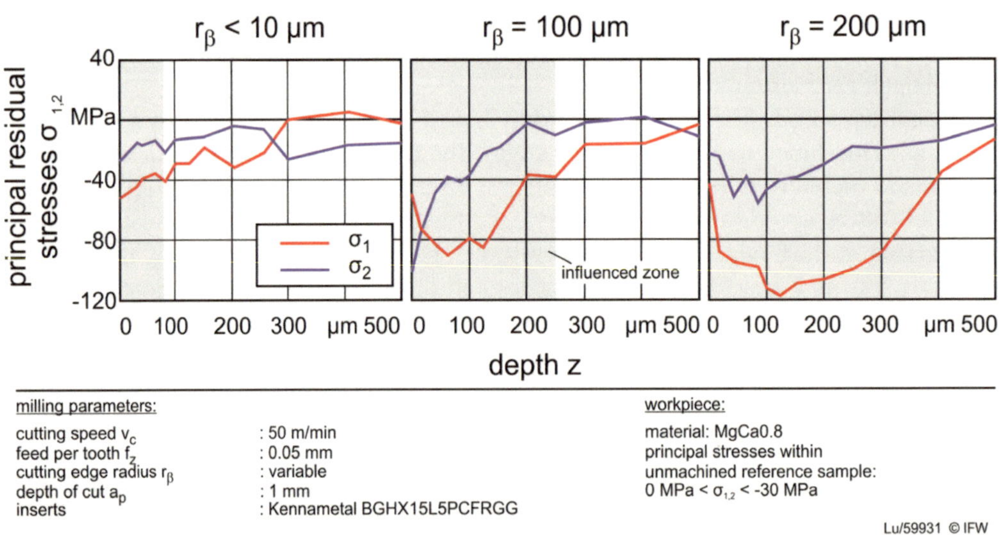


Fig. 5. Influenced subsurface after milling with modified tools

### 3.1.4 Correlation of results of studies on turning and milling

To correlate the results of the studies on turning and milling, test series are carried out, in which similar process parameters are adjusted. Therefore similar chip thicknesses result for both machining methods. That is especially, because the residual stress is measured in the center of the milling path, where the chip thickness equals the feed per tooth. In both processes extruded materials (Ø = 20 mm) are used, which are turned to a diameter of Ø = 18 mm respectively milled into a half cylinder. Furthermore, turning and milling tools are prepared with similar cutting edge roundings (sharp $r_\beta < 10$ µm, $r_\beta = 100$ µm, $r_\beta = 200$ µm). Therefore, analogical relations exist between chip thickness and cutting edge radii. The results of the residual stress measurements illustrated in Figure 6 show similar tendencies in the depth profiles of the workpieces after turning and milling. There is only a minor modification of the subsurface for sharp tools. With increased rounding of the cutting edge squeezing effects and therefore plastic deformation of the material close to the surface increases, which induces residual compressive stress into the subsurface. This effect becomes even more significant in milling at increased cutting edge radii.

Comparing the depth of impact of both processes, significant differences are detectable. In turning a depth influence of approx. 100 µm can be achieved, which cannot be increased by doubling the radius of the cutting edge. In milling with modified tools of $r_\beta = 100$ µm a depth of influence of up to approx. 200 µm can be achieved and for $r_\beta = 200$ µm a depth of up to approx. 400 µm.

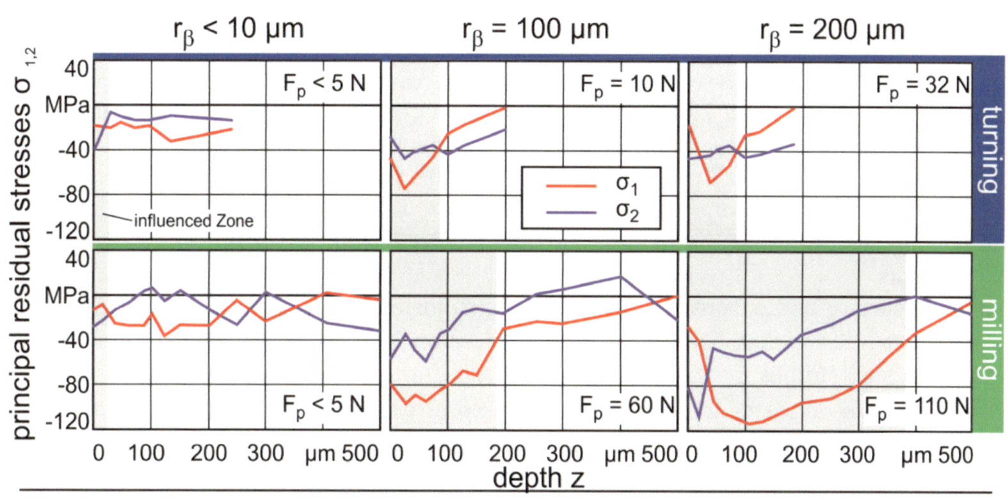


| process parameters: | turning / milling | workpiece: |
|---|---|---|
| cutting speed $v_c$ | : 50 m/min / 50 m/min | material: MgCa0.8 |
| feed f / feed per tooth $f_z$ | : 0.05 mm / 0.05 mm | principal stresses within |
| cutting edge radius $r_\beta$ | : variable | unmachined reference sample: |
| depth of cut $a_p$ | : 0.1 mm / 0.1 mm | 0 MPa < $\sigma_{1,2}$ < -30 MPa |
| inserts | : Kennametal CPGT09T308LF / Kennametal BGHX15L5PCFRGG | |

Lu/59931 © IFW

Fig. 6. Comparison of the subsurface modification after turning and milling

The forces in passive direction can be considered as the main cause of this effect. These forces ($F_p$) directed normally toward the surface of the workpiece are on a higher level for milling than for turning. The continuous chip generation in turning with rounded tools obviously leads to a jam of material in front of the cutting edge. The jammed material does

not flow neither over the cutting edge, nor below and therefore acts like a built-up edge. The negative effective rake angle resulting form the rounding of the cutting edge therefore does not come into effect and only a minor increase of the passive force is detected. The milling process shows that at the applied low cutting speed no ordinary built-up edge is generated. In discontinuous cutting the edge cannot form from jammed material because the height of the point of stagnation varies with the varying undeformed chip thickness and, in addition to this, the tool periodically gets out of contact with the work material.
The discontinuous milling process also lies on a lower temperature level due to the cutting edge cooling in the interruption of cut. Furthermore, the machined surface and therefore the volume for a similar feed distance is larger in the turning process (proportional to the circumference U = 56.5 mm at Ø = 18 mm) than in the milling process (width of milling path b = 20 mm). The thermal influences inducing tensile residual stress are therefore lower in the milling process and the mechanical influences inducing compressive residual stress dominate. This is also confirmed by the residual stress of the surface, which hardly exceed -40 MPa in turning while in milling up to -80 MPa are achieved.

### 3.1.5 In vitro corrosion studies of the samples mechanically machined

After analyzing the surface modifications and the depth effects of the subsurface modifications caused by machining processes, in vitro corrosion studies are performed, to examine the relation between these modifications and the impact on the corrosion behavior, especially on the corrosion profile over the time. It was observed how the corrosion characteristics of implants with modified subsurfaces change over time with regard to environmental influences. This provides information on the period of time in which the modified subsurface has an influence on the corrosion and degradation behavior of the implant ($t_{lim}$ in Figure 2). To be able to determine the influence of the modification of the surface and subsurface on the corrosion behavior, the corrosion progress of different subsurface modifications is monitored time-resolved and analyzed in the corrosion experiments.
Different surface and subsurface properties are adjusted on similar sample geometries in order to analyze the time behavior of the degradation induced by the machining. In the in vitro corrosion studies in physiological NaCl solution (0.9 wt% NaCl in deionized water) carried out. The mass of the corroding magnesium can be determined via continuously monitoring volume of the generated hydrogen released during the oxidation of magnesium. The samples are produced by means of the chip removing and non-chip removing processes described earlier.
The studies on the machining influence on different magnesium materials revealed a major difference in the ability to influence different alloys, considering the corrosion. While the surface and subsurface properties resulting from machining are similar for both materials MgCa0.8 and MgCa3.0 (see above) the corrosion behavior of the two differs significantly. Only a minor influence of the mechanical processing on the resulting corrosion kinetics was revealed for MgCa0.8. Figure 7 and Figure 8 show that a very similar corrosion progress was revealed for turning and deep rolling of MgCa0.8 despite very different surface and subsurface properties. The μ-CT analyzes are performed in cooperation with the IW.

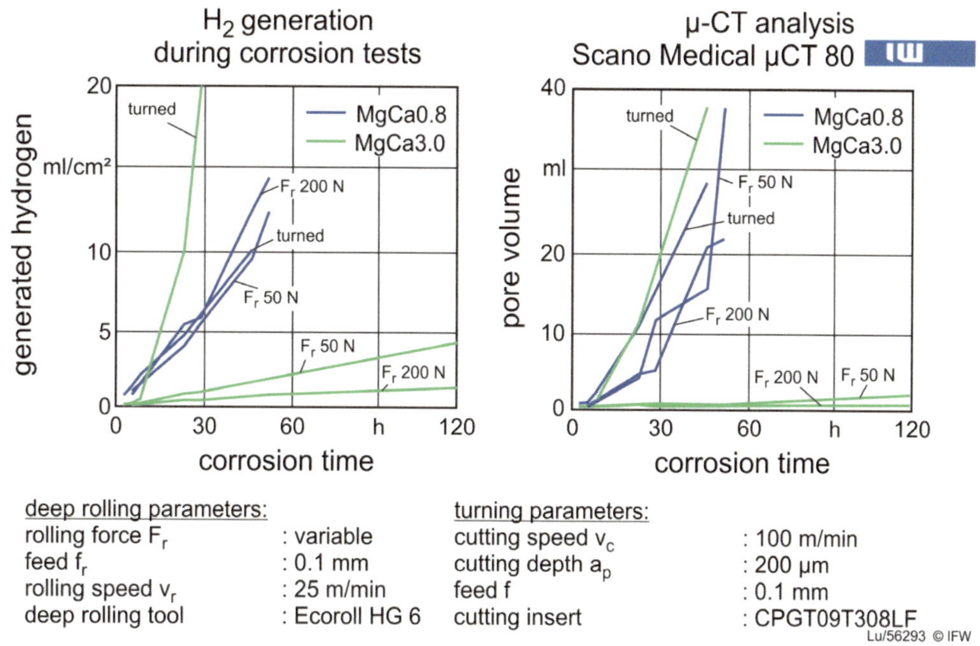


Fig. 7. Corrosion kinetics of turned and deep rolled samples of different material

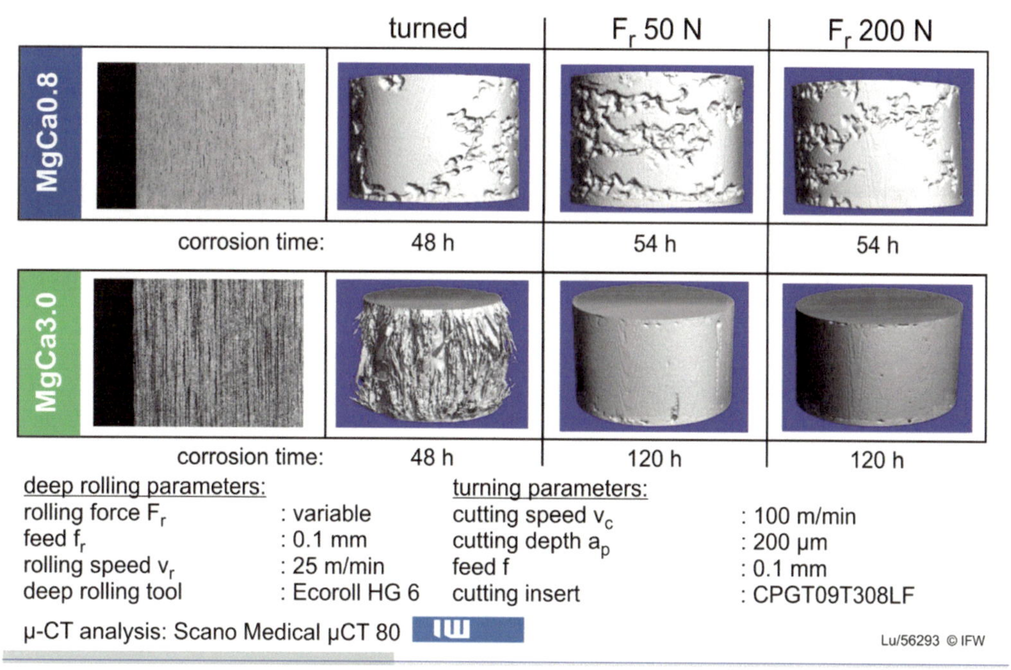


Fig. 8. Corrosion morphology of turned and deep rolled samples of different material

The records of the release of hydrogen during the corrosion (Figure 7) as well as the evaluation in the μ-CT of the samples that were analyzed after a certain time (Figure 8, top)

show that no significant differences concerning the corrosion behavior of MgCa0.8 are detectable for the machining methods and process parameters applied. The corrosion rates are at a similar level. The alloy MgCa3.0, though, showed a significantly different corrosion behavior. This alloy shows a strong sensitivity of the corrosion behavior towards the modification of the subsurface. The corrosion progresses monitored by means of the hydrogen generation over time (Figure 7) as well as the samples studied after certain corrosion times in the µ-CT (Figure 8 bottom) show a strong influence of the mechanical processing. A significant reduction of the corrosion rate can be achieved by means of deep rolling. A difference in the corrosion rates of MgCa3.0 in turned and deep rolled samples by a factor of up to 100 is proved by the corrosion studies. The turned MgCa3.0 samples are attacked very strongly and their surface is undermined and then even bursting by the corrosion attack. The corrosion in deep rolled samples progresses delayed and very homogenously. In contrast to the homogenous structure of MgCa0.8 in which the calcium is dissolved interstitially, in MgCa0.3 the phase of the less noble $Mg_2Ca$ is attacked stronger. The deep rolling process obviously has an effect of closing superficial mirco-pores and additionally the residual compressive stresses induced avoid a superficial formation and propagation of cracks. This prevents the corrosion front from progression toward the center of the workpiece and the subversion of the surface. This in turn results in an overall homogenous corrosion attack. Thus, besides the basic corrosion rate of an applied alloy not only the surface and subsurface properties are decisive for influencing the corrosion rate. But also the characteristics of the alloy regarding its micro structural composition and crystallization are of great importance in the achievable modification bandwidth. Therefore it can be distinguished between materials with a corrosion behavior sensitive to machining and those that can be modified only very limitedly.

### 3.2 Clinical in vivo trials

The connection between the surface and subsurface properties and the in vivo degradation behavior as well as the resulting biomechanical loss of function of implant demonstrators is investigated in the field "Clinical in vivo trials". Therefore, the degradation kinetics of magnesium implants of different surfaces, subsurfaces and geometries were studied in experiments on rabbits.

#### 3.2.1 Studies on cylindrical implant demonstrators

To study the influence of the mechanical processing on the surface and subsurface of cylindrical implant demonstrators and their corresponding corrosion behavior, a magnesium calcium alloy with a calcium content of 0.8 % is chosen. This shows good results in both, in vitro and in vivo pilot studies. Smooth and sandblasted implants as well as threaded cylinders out of MgCa0.8 are studied. Besides the testing of the in vivo corrosion characteristics also the biocompatibility in the implant-bone bonding and the osseointegration are tested. The cylinders are implanted into the cortico-spongy transition area of the medial condyle of femur and remained for evaluation periods of 3 and 6 months. Besides clinical and radiological monitoring µ-CT scans are made to evaluate the existence of bone formation, gas accumulations and the state of degradation of the implants at the end of the trial period and after explantation of the distale fermur. Then, the non-decalcified bone-implant-compound is embedded in artificial resin and analyzed histological, after the preperation of microtome sections. Further, the histological cross sections of the cylinders are studied by means of an SEM, including EDX.

The investigations verify results from pilot studies: the sandblasted cylinders show the highest level of degradation and a n early loss of shape, the smooth cylinders degrade very slowly and homogeneously and the threaded cylinders show local corrosion attacks in the crest areas of the thread (Figure 9).

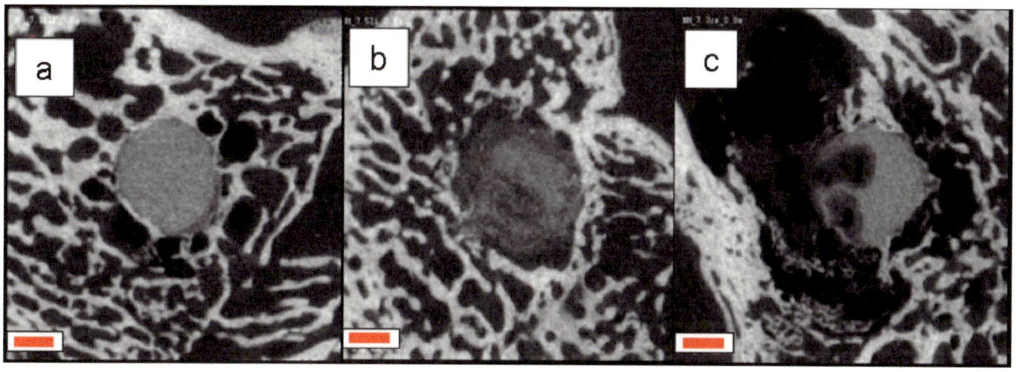


Fig. 9. μ-CT scans of a smooth cylinder (a), sandblasted cylinder (b) and threaded cylinder (c) in the condyle of femur six months after implantation in the rabbit

The generation of gas as well as the biocompatibility correlates with the state of degradation. The faster degrading sandblasted cylinders show the strongest generation of gas with clearly visible bubbles under the skin in close vicinity to the implant in 50 % of the cases. Lymphocytes, plasma cells and foreign body cells, as signs of the immune response of the organism, appear most frequently in the tissue around the sandblasted implants. Smooth cylinders do not show any gas accumulations and the cellular reactions are minimal. The gas accumulation on the threaded cylinders is moderate. It was mostly only diffuse and small accumulations that appeared in the tissue. All of the implants led to the formation of small patches of cartilaginous tissue around the implants, which can be seen as an early stage of the remodeling of the bone. The further development of these patches has to be analyzed in studies over longer implantation periods. A slower and more consistent degradation as seen with the smooth implants is more favorable because the reaction of the surrounding tissue is the lowest.

### 3.2.2 Studies of screw geometries

The osseointegration of screw geometries after different implantation periods (2, 4, 6 and 8 weeks) is tested. Due to the screw dimensions, the cortico-spongy transition area of the distal femur proved to be inappropriate in pre-tests. Therefore the screws are implanted in the lateral tibia of the rabbit at the level of the insertion site of the fibula. Screws out of the magnesium alloy MgCa0.8 and surgical steel 316L are utilized. The production of the screw occurs following the in vitro tests (comparable thread pitch and flank width) to be able to compare in vitro and in vivo results. During the testing period, clinical, radiological and in vivo μ-CT examinations are carried out, especially to evaluate the degree of bony and emphysematous formations. Even during the trials, an evaluation on the degradation of the Mg screws could be carried out via of the μ-CT data.

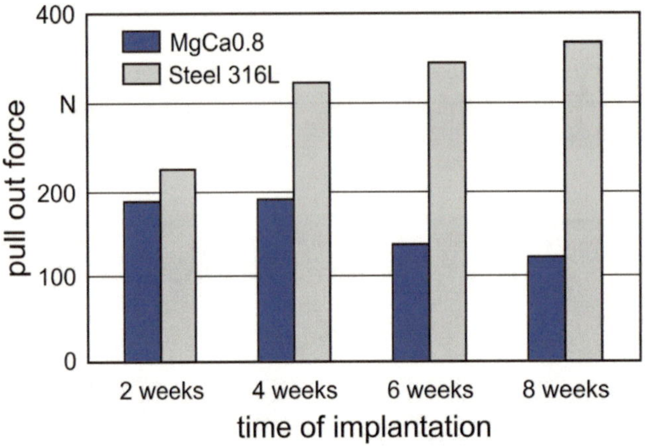


| time of implantation | pull out force MgCa0.8 | pull out force Steel 316L |
|---|---|---|
| 2 weeks | 186 N | 225 N |
| 4 weeks | 188 N | 322 N |
| 6 weeks | 136 N | 346 N |
| 8 weeks | 123 N | 368 N |



Fig. 10. Tensile forces of resorbable MgCa0.8 and permanent 316L steel screws after different implementation periods

At the end of the evaluation period the strength of the implant-bone compound is tested by biomechanical examinations: to evaluate the degree of osseointegration and biodegradation pull-out tests are carried out (Figure 10). A special test set-up was established that allows to pull out the screw demonstrators with geometries from the in vivo studies as well as from the in vitro studies for comparison. The pull-out tests show no significant difference between the maximum pull-out forces of MgCa0.8 and surgical steel after an implantation period of two weeks. The maximum pull-out force of MgCa0.8 hardly decreased within the first four weeks after the implantation. Only in the long term, a continuous decreasing of the pull-out forces is detected. In contrast to that, the pull-out force of surgical steel continuously increases during the period of implantation. Therefore, significantly higher pull-out forces are measured for the surgical steel after four, six and eight weeks. Eventually, the screws were analyzed in toto and, after embedding in artificial resin, in thick-sections by means of SEM and EDX.

The muscles situated above the screw head (M. tibialis cranialis) are examined histological to evaluate the biocompatibility of the screws. The prepared paraffin sections were stained with H&E staining and analyzed (Figure 11). The histological analysis of the peri-implant tissue shows similar results for both implant materials, which proves the formation of a fibrotic layer between the screw head and the covering muscles. Macrophages, foreign body cells and heterophile granulocytes as well as small patches of necrotic tissue are identified. The analysis focuses on the evaluation of the degree of fibrosis (thickness of fibrotic layer), necrosis and the cellular infiltration (amount of macrophages, foreign body cells, heterophile granulocytes) and is carried out by means of semi-quantitative scoring (non-existent, insignificant, low-grade, medium-grade, high-grade, very high-grade). In the group of steel implants the thickness of the connective tissue, the degree of cellular infiltration and the necrosis decreases continuously. A decrease of the parameters analyzes also appears in the MgCa0.8 group, however only the first four to six weeks after surgery. At the end of the implantation period the parameters analyzed increase.

Immunhistochemical staining is applied to show CD3 and B-CD79α, by which small amounts of T- and B-lymphocytes are detectable in both groups (MgCa0.8 and steel 316L). In comparison to steel, fewer T-lymphocytes and more B-lymphocytes are traced in the MgCa0.8 group.

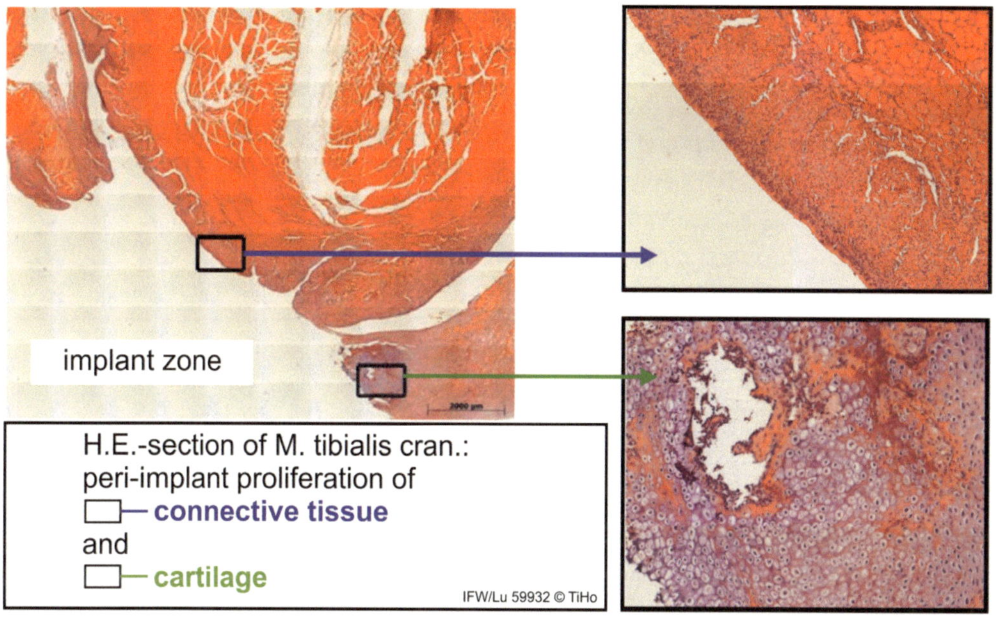


Fig. 11. Histological scans of a muscle section after the implantation of an MgCa0.8 screw

### 3.3 Biomechanics

In the field of "Biomechanics", the influence of modified surface and subsurface conditions on the in vitro corrosion behavior under different loads is analyzed and the biomechanical loss of function due to degradation of the material is modeled. The performance of the implant demonstrators are tested in vitro under dynamic stresses. Different stresses of the samples are reproduced in biomechanical test setups. The analysis of components in graded states of corrosion provides information on the loss of function as a result of geometry changes and mass loss due to corrosion.

A newly developed and established corrosion test stand outlines the loss of function depending on the corrosion progress. The corrosion stand can be divided into three sub-units. The flow velocity is adjustable by a peristaltic pump, the first unit. The thermostat as a further unit indirectly warms up the corrosion solution, which is led through a heat bath. The temperature of the solution is checked by means of a thermometer before it flows into the corrosion chamber. The corrosion chamber is the third unit, which is adjustable for each demonstrator-geometry. The chamber design ensures a laminar flow within the corrosion chamber. The significant factors of corrosion such as flow velocity, temperature of the solution and its composition are adjustable in the corrosion stand. Hence, the influences of the modified surface and subsurface conditions on the in vitro corrosion behavior under static and dynamic stresses and the biomechanical loss of function due to degradation of the material can be determined.

#### 3.3.1 Static stress with and without corrosion

The aim of the investigation is to determine the maximum pull-out force without influence of the corrosive medium. MgCa0.8 screws with an outer diameter of 4 mm, a core diameter of 3 mm and a thread pitch of 1 mm are applied in this test. They are inserted into artificial bone out of polyurethane (SYNBONE AG, Neugutstrasse 4, CH- 7208 Malans). Density of cortical plate: 0.72 g/cm³ +/ 5%; Density of cancellous bone: 37 kg/m³ (DIN EN ISO 845). By means of a biomechanical test device (Bionix-MTS 14000, Technology Drive Eden Prairie, MN USA) the screws are exposed to an axial pull-out force that was linearly increased until the compound between screw and artificial bone failed.

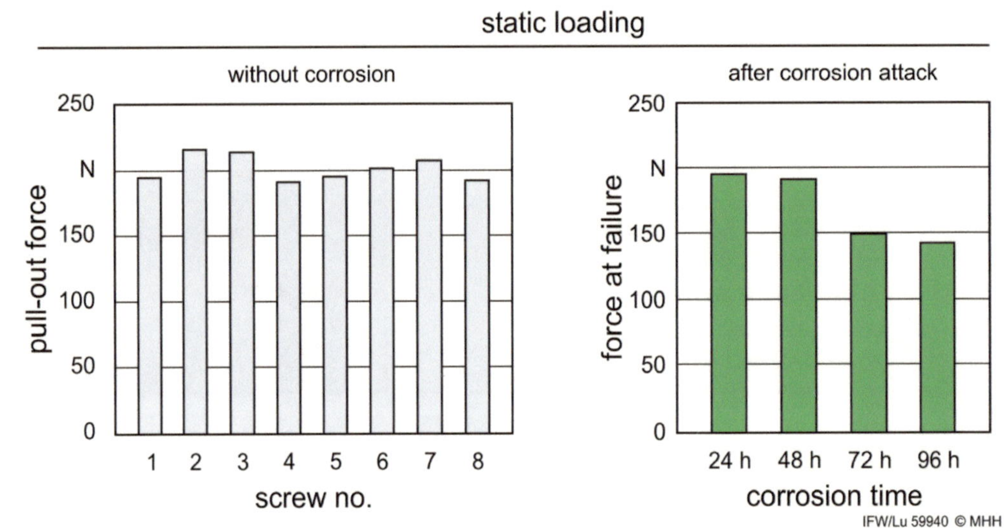


Fig. 12. Pull out forces of screw demonstrators (left) and mean value of the force at failure after 24h, 48h, 72h, 96h of corrosion for groups of 8 screws (right)

In each test series eight screws are used. The results for the demonstrators without corrosion show a reproducible bone-screw fixation at a low standard deviation. This data serves as a reference value for the determination of the failure load of the bone-screw compound at different states of corrosion. The left diagram in Figure 12 shows an average pull-out force of 201.5 N at a standard deviation of 9.3 N that appeared in the pull-out tests.
To detect the remaining maximum pull-out force after corrosion (24 h, 48 h, 72 h and 96 h) MgCa0.8 screws of an outer diameter of 4 mm, a core diameter of 3 mm and a thread pitch of 1 mm are applied. In the test stand they are inserted into the artificial bone out of polyurethane. The flow rate of the used HANK's solution is set to 2 ml/h at an adjusted temperature of 37° C. After fixed intervals, the screws were stressed until failure by means of the MTS test device. For each corrosion period eight screws were tested: 8/24 h; 8/28 h; 8/72 h; 8/96 h. The maximum tensile force decreases under the influence of corrosion over time (see Figure 12, right) and the preliminary assumptions were therefore verified.

### 3.4 Correlation of biomechanical results in vitro / in vivo

For the design of implants and the prediction of the implant behavior in vitro results have to be correlated with those from the clinical in vivo experiments. Figure 13 shows similar tendencies in the development of biomechanical characteristics over time in both environments. A continuous function loss of the screw demonstrators is proved after corrosion in vitro and degradation in vivo, respectively. Furthermore, a strong scaling of the behavior over time becomes obvious. While the decrease of the tensile force of approx. 30 % occurs within 100 h in vitro, a similar function loss in vivo only occurs after about 1000 h. The time profiles of the investigated biomechanical characteristics therefore are accelerated ten times approximately under in vitro conditions.

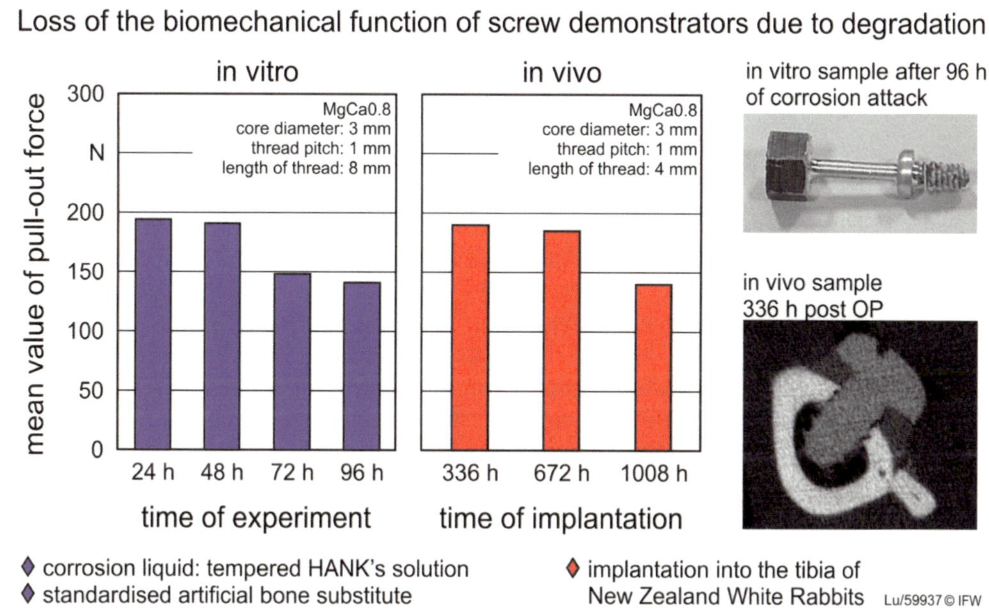


Fig. 13. Comparison of the biomechanical loss of function in vivo and in vitro

## 4. Conclusion and outlook

Revision surgeries on implants are a significant cost driver in public health care. Avoiding the material removal surgeries, resorbable implants out of magnesium therefore offer a great potential for cost reduction. Additionally, the patient is not unnecessarily put to risk.

The works presented focus on the idea of "intelligent osteosynthesis" in which implants with a specific degradation profile are subjected to a stability loss, which is adjusted to the increasing stability of the healing bone. The presented results show that specific surface and subsurface modifications can be achieved by means of adapted machining processes and for suitable alloys the resulting corrosion behavior can be adjusted in vitro and the degradation profile in vivo, respectively. Biomechanical tests prove a continuous stability loss due to degradation in vitro and in vivo.

In the future, the application of resorbable implants enables new areas of indication by the development of implant designs adapted to the material and the expected degradation as well as by new methods for the production including an adapted mechanical processing of these magnesium implants. Producing these implants in industrial production processes that guarantee specific implant properties without exceeding the production costs of currently used implants will be challenging. To enable the clinical application and achieve the marketability for load bearing resorbable magnesium implants, future tasks will focus the development and adaptation of the production technology and the adjustment of specific implant properties corresponding with the geometrical implant design. At sufficient stability, fatigue strength, resulting in a stable osteosynthesis until completion of the bone healing, the field of application of load bearing resorbable implants can be made accessible.

## 5. Acknowledgments

The investigations described in this paper were funded by the German Research Foundation (DFG) within the subproject R4 of the collaborative research center SFB 599 "Sustainable Bioresorbable and Permanent Implants of Metallic and Ceramic Materials".

## 6. Reference

Anker, C. J.; Holdridge, S. P., et al. (2005). Ultraporous beta-tricalcium phosphate is well incorporated in small cavitary defects. *Clin. Orthop. Relat. Res.*, 434:251-257

ASTM-Standard F543. (2002). *Annual Book of ASTM Standards,* Philadelphia, Pennsylvania/USA: American Society for Testing and Materials

Brinksmeier, E., Garbrecht, M., Meyer, D., & Dong, J. (2007) Surface hardening by strain induced martensitic transformation. *Production Engineering,* (2)2:109-116

Broszeit, E., & Adelmann, J. (1996). *Schwingfestigkeitssteigerung durch Festwalzen – Grundlagen und Anwendung*. Broszeit, E.; Steindorf, H. (Eds.). *Mechanische Oberflächenbehandlung*. DGM Informationsgesellschaft Verlag, Oberursel, pp. 63-82

Carano, A., Lonardo, P., Velo, S., & Incorvati, C. (2005). Mechanical properties of three different commercially available miniscrews for skeletal anchorage. *Prog Orthod.,* 6(1):82-97

Daniels, A. U., Chang, M. K., et al. (1990). Mechanical properties of biodegradable polymers and composites proposed for internal fixation of bone. *J. Appl. Biomater,* 1:57-78

DeCoster, T. A., & Heetderks, D. B., et al. (1990). Optimizing bone screw pullout force. *Journal of Orthopaedic Trauma,* 4(2):169–174

Denkena, B., Alpers, P., Weinert, K., & Hammer, N. (2006). Influence of Cutting and Non-Cutting Processes on the Corrosion Behavior and the Mechanical Properties of Magnesium Alloys. *Production Engineering,* XIII(1):1-4

Denkena, B., Podolsky, C., van der Meer, M., & Lucas, A. (2006). Machining of Permanent and Bioresorbable Implants Made of Ceramics and Magnesium. *Production Engineering,* XIII(1):13-18

Denkena, B., Boehnke, D., & de León, L. (2008). Machining induced residual stress in structural aluminum parts. *Production Engineering,* 2(3):247-253

Denkena, B., & Lucas, A. (2007). Biocompatible magnesium alloys as absorbable implant materials - Adjusted surface and subsurface properties by machining processes. *Annals of the CIRP,* 56(1):113-118

Gotz, H. E., Muller, M., et al. (2004). Effects of surface finish on the osseointegration of lasertreated titanium alloy implants. *Biomaterials,* 25(18):4057-64

Ignatius, A. A., Augat, P., Hollstein, E., Schorlemmer, S., Peraus, M., Pokinskyj, P., & Claes, L. (2005). Biocompatibility and functionality of the degradable polymer alkylene bis(dilactoyl)-methacrylate for screw augmentation in vivo. *J Biomed Mater Res. B Appl Biomater,* 75(1):128-136

Ignatius, A. A., Betz, O., et al. (2001). In vivo investigations on composites made of Resorbable ceramics and poly(lactide) used as bone graft substitutes. *J. Biomed. Mater. Res.* 58:701-709

Ignatius, A. A., Ohnmacht, M., et al. (2001) A composite polymer/tricalcium phosphat membrane for guided bone regeneration in maxillofacial surgery. *J. Biomed. Mater. Res.* 58:564-569

Inceoglu, S., Ferrara, L., & McLain, R. F. (2004). Pedicle screw fixation strength: pullout versus insertional torque. *Spine J.*, 4(5):513-518,

Kim, Y. H., Koak, J. Y., et al. (2003). A histomorphometric analysis of the effects of various surface treatment methods on osseointegration. *Int J Oral Maxillofac Implants,* 18(3):349-356

Kloos, K. H., Adelmann, J., et al. (1988). Oberflächen- und Randschichteinflüsse auf die Schwingfestigkeitseigenschaften. *VDI-Berichte,* 661:215-245

Mueller, W. D.; Gross, U.; et al. (2003). Evaluation of the interface between bone and titanium surfaces being blasted by aluminium oxide or bioceramic particles. *Clin Oral Implants Res.,* 14(3):349-356

Nasatzky, E., Gultchin, J., et al. (2003). The role of surface roughness in promoting osseointegration. *Refuat Hapeh Vehashinayim,* 20(3):8-19, 98

Niemeyer, M., Windhagen, H., et al. (2001). Magnesiumlegierungen als degradabler Werkstoff für die medizinische Anwendung. *Biomedizinische Technik, Ergänzungsband,* 46(1):238-239

Nowag, L., Sölter, J., & Brinksmeier, E. (2007). Influence of turning parameters on distortion of bearing rings. *Production Engineering,* 1(2):135-139

Nüstedt, H. (1996). *Beitrag zum Ermüdungsverhalten superleichter Magnesium-Lithium-Basislegierungen,* Dr.-Ing. Dissertation, Universität Hannover

Sakar, M., Bakker, F. C., et al. (1999). Biobon - ein Calciumphosphat-Zement zur Auffüllungknöcherner Defekte. Ergebnisse der klinischen Pilotstudie mit 49 Patienten. *Unfallchirurg,* 275:238-239

Schneider, G. B., Perinpanayagam, H., et al. (2003). Implant surface roughness affects osteoblast gene expression. *J Dent Res.,* 82(5):372-376

Scholtes, B., & Vöhringer, O. (1996). *Grundlagen der mechanischen Oberflächenbehandlung;*

Broszeit, E., & Steindorf, H. (Eds.). *Mechanische Oberflächenbehandlung,* DGM Informationsgesellschaft Verlag, Oberursel, pp. 3-20

Scholtes, B., & Zinn, W. (1997). *Mechanische Oberflächenbehandlung von Leichtbauwerkstoffen,* Frieling, E., Martin, H., & Tikal, F. (Eds.). *Neue Ansätze für innovative Produktionsprozesse,* Kassel University Press, Kassel, pp. 376-383

Seebeck, J., Goldhahn, J., Morlock, M. M., & Schneider, E. (2005). Mechanical behavior of screws in normal and osteoporotic bone. *Osteoporos Int.,* 16(2):107-111

Seitz, J. M., Wulf E., Freytag, P., Bormann, D., & Bach, F. W. (2010). The Manufacture of Resorbable Suture Material from Magnesium. *Advanced Engineering Materials,* 12(11):1099-105

Shalabi, M. M., Gortemaker, A., et al. (2006). Implant surface roughness and bone healing: a systematic review. *J Dent Res.,* 2006 85(6):496-500

Taylor, M. S., Daniels, A. U., et al. (1994). Six bioabsorbable polymers: in vitro acute toxicity of accumulated degradation products. *J. Appl. Biomater.,* 5:151-157

Tomac, N., & Tonnessen, K. (1991). Formation of Flank Build-Up in Cutting Magnesium Alloys, *Annals of the CIRP*, 40(1):79-82

Tönshoff, H. K., Haferkamp, H., et al. (2001). *Einfluss des Festwalzens auf die Oberflächen- und Randzoneneigenschaften der Magnesiumlegierung AZ31 und AZ91,* Sonderband Magnesiumtechnologie. Materialwissenschaft und Werkstofftechnik

Weiler, A. M., Hoffmann, R. F. G., et al. (2000). Biodegradable implants in sports medicine. The biological base. *Arthroscopy,* 16:305-321

Weiler, A. M., Windhagen, H. J., et al. (1998). Biodegradable interference screw fixation exhibits pull-out force and stiffness similar to titanium screws. *Am. J. Sports Med.,* 26:119-126White, K. K., Oka, R., Mahar, A. T., Lowry, A., & Garfin, S. R. (2006). Pullout strength of *thoracic pedicle screw instrumentation: comparison of the transpedicular and extrapedicular techniques.* Spine, *31(12):E355-E358*

Winkler, J. (2000). *Herstellung rotationssymmetrischer Funktionsflächen aus Magnesiumwerkstoffen durch Drehen und Festwalzen,* Dr.-Ing. Dissertation, Universität Hannover

Wintermantel, E., & Ha, S. W. (1996). *Biokompatible Werkstoffe und Bauweisen,* Springer, Berlin, Heidelberg, New York

Wittenberg, J. M., Wittenberg, R. H., et al. (1991). Biomechanical properties of resorbable poly-Llactide plates and screws: a comparison with traditional systems. *J. Oral Maxillofac. Surg.,* 49:512-516